DECENTRALISATION AND CHANGE IN CONTEMPORARY FRANCE

Decentralisation and Change in Contemporary France

MICHAEL KEATING
Lecturer in Politics, University of Strathclyde
and
PAUL HAINSWORTH
Lecturer in Politics, University of Ulster

Gower

© M. Keating and P. Hainsworth, 1986.

All rights reserved. No part of this publication may be reproduced, stored in a retrieval system, or transmitted in any form or by any means, electronic, mechanical, photocopying, recording, or otherwise, without the prior permission of Gower Publishing Company Limited.

Published by
Gower Publishing Company Limited
Gower House, Croft Road, Aldershot, Hants GU11 3HR
England

Gower Publishing Company
Old Post Road, Brookfield, Vermont 05036, USA

ISBN 0 566 05206 7

Printed and bound in Great Britain by
Paradigm Print, Gateshead, Tyne and Wear

Contents

Preface   3

1. Perspectives on the State in France.   5
2. Before Mitterrand.   15
3. Regionalism and Micronationalism in Mainland France.   34
4. Decentralisation and the Parti Socialiste.   53
5. The Programme in Government.   70
6. The Change in Corsica.   92
7. Decentralisation and the Economy.   109
8. Conclusion.   129
Bibliography   136
Index   142

Preface

Decentralisation of power was, according to former Interior Minister Gaston Defferre, to be the 'grand affaire' of the Mitterrand Presidency. Centuries of centralised control were to be reversed and the French state, in Mitterrand's own words, given back to the people. There was to be a 'new citizenship' for the individual, recognition for submerged ethnic and cultural minorities and a new division of economic power. For some, indeed, decentralisation was part of the rupture with capitalism and the creation of a 'third way' between the market economies of the west and the bureaucratic socialism of the east. These were ambitious aims and, while the new Socialist Government of 1981 lost little time in starting on the programme, its realisation has, perhaps unsurprisingly, been slower and less radical than anticipated. This book traces the implementation of the decentralisation programme and makes a provisional judgement on its success.

Of course, the need for decentralisation has been a theme in proposals for the reform of French government for many years. Indeed, the complexity of the subject and the differing interpretations of the structure of the French state have given rise to a considerable literature. In our first chapter, we review the major interpretations of French territorial government to identify the main issues at which reform proposals have been addressed. Then we examine the reform efforts of previous governments. Chapters three and four set the context for the Socialist Government's preoccupation with decentralisation, tracing the growth of regionalism and micronationalism in France, with case-studies of Brittany and Languedoc, and showing the impact of this on the Parti Socialiste at a time of partisan realignment in the 1970s. At the same time, we note the multiplicity of strands within the Parti Socialiste, including the traditional territorial notables and a strong surviving 'jacobin' element. Chapter five examines the legislation and implementation of the decentralisation programme after 1981 and the impact of this on power relations, while chapter seven considers the statut particulier for Corsica. Chapter six looks at decentralisation and economic policy, a key item in the Socialists' early vision. The outcome of the programme will not be apparent for some years, as new power relationships are established but in the conclusion we present a preliminary assessment.

We have received help and support from many people in the course of our research and would like to thank particularly Jacques Leruez of the Fondation Nationale des Sciences Politiques, Paris, Albert Mabileau and his colleagues at the CERVL, Bordeaux; Yves Meny of the University of Paris and Paul Allies and colleagues at the University of Montpellier. Financial assistance was given by the Nuffield Foundation, the Leverhulme Trust and the Economic and Social Research Council. Our thanks go, too, to Alison Robinson and Grace Hunter who typed the final version.

Michael Keating
Paul Hainsworth

November, 1985.

1 Perspectives on the state in France

THE JACOBIN MYTH

Centralisation of power and the tradition of 'one and indivisible republic' are often seen as the hallmarks of the French state. The 'napoleonic' system of administration is taken as synonymous with order, unity and uniformity. In fact, the centralist tradition long predates Napoleon, relecting the struggle of French rulers over centuries to establish the external boundaries and internal cohesion of the collection of territories now known as France. Nor does it draw only on the authoritarian inheritance of bonapartism or monarchy. The French revolution, asserting the sovereignty of the people, reasserted that of the nation. If national sovereignty was now identical with the sovereignty of the French people, then that people must be united in the face of the world. The jacobins of the revolutionary period may not have been the ferocious centralisers of subsequent mythology (Hayward, 1983); but since the nineteenth century the term 'jacobinism' has often been synonymous with the defence of an ideal of a uniform, centralised, sovereign and democratic France. The associations of centralisation with the revolutionary values of liberty, equality, fraternity and democracy is crucial to the maintenance of the jacobin myth, in sharp contrast to Britain and the United Staes, where the liberal tradition associates democracy and liberty with decentralisation

(Mény, 1974). Local particularisms and dispersed power provide shelter for reactionary elements, threatening to the unity of the state and the sovereignty of the people. Civil equality requires that all citizens face the state directly, with no intermediate institutions or competing foci of loyalty.

Centralisation is intimately bound up with nationalism, the idea that the French people are a natural entity to which one's first loyalty is due. The transfer of sovereignty from monarch to people in the course of the revolution allowed post-revolutionary democrats to take over the centralising ideology of the monarchy and to complete its centralising and unifying mission, most notably during the Third Republic (Weber, 1977). Third republic schoolteachers taught of the long struggle of the 'French nation' to realise itself (Lebesque, 1970) and the Napoleonic system of administration was defended against all comers. The centralising efforts of nineteenth-century republicans were fuelled by their view of the provinces as a haven of reaction, clericalism and royalism, a view sustained by the attitudes of many contemporary regionalists. So, while liberals saw centralisation as vital both to the defence of the republic and to such reforms as the Ferry laws establishing universal secular education and breaking the power of the clergy, anti-republicans saw in it the defence of old tradiations and institutions against the godless state. Centralisation thus became a fundamental belief of many progressives as well as of the authoritarian and bonapartist right. Finally, jacobinism undoubtedly had strong appeal to the French intellectual penchant for order and rationality and a related belief that only centralised and uniform action is rational (Grémion, 1976).

We have spoken deliberately of the jacobin 'myth' to indicate a set of beliefs as to the state of the world which may or may not be true but act as a mobilising ideology. Both jacobins and girondins - the inheritors of the revolutions's defeated decentralist wing - have tended to regard the jacobin vision as a reality, the one to defend it, the other to attack it. Certainly, the formal features of the French state lend support to the centralist view. The basic units of local government are the 36,000 communes, dating from Napoleonic times but based on the pre-revolutionary parishes. Each is governed by a mayor, directly elected since 1884 and an elected council. As well as being the elected representative of the commune, the mayor is an official of the state. Communes vary enormously in size and resources, from the city of Marseille, with a population in the region of a million, to small rural communities with a dozen or so permanent inhabitants. Apart from Paris, which is both a département and a commune, and Marseille, cities themselves tend to be fragmented into several communes, further weakening their capacity for purposive action. It is true that some urban communes are grouped into communautes urbaines and other some together in various forms of single and multi-purpose syndicates, but these are essentially forms of intercommunal co-operation, not bringing into question the independence or integrity of the basic communal unit. Until the recent reforms, all communal decisions were subject to the tutelle, or control, of the Prefect, who could veto actions either on legal or on general political grounds. Most of the rural communes are in any case very small and poor and dependent on central government agencies for advice and guidance as well as material support.

The next level are the 96 départements, a Napoleonic creation, with elected councils (the conseil général) but, until the recent reforms, no executive of their own. Instead, the Prefect acted as executive and prepared the council's budget which was voted at its twice-yearly meeting. The administration of the département was undertaken largely by the prefectoral services or by the departmental agencies of the central ministries. As a post-revolutionary creation, the département is invested with a great deal of importance by republicans but it is widely thought to be an outdated unit. Its functions have been strictly limited at least until the recent reforms and political life has been debilitated by the archaic electoral system with its inbuilt rural bias.

Since 1972, there have been indirectly-elected regional councils, again without their own executive until recently and with very limited financial and administrative power. Their establishment was less a concession to pressures for regional decentralisation than a means of modernising the administration system at minimum political cost. They were not encouraged to establish their own services - and indeed had to resort to various subterfuges to do so - but were expected mainly to support the interventions of other agencies.

While, as we note in the next chapter, many city communes and départements have built up their own services, most services remain in the hands of central ministries which can impose stringent technical conditions (the tutelle technique) on local councils before approving works. The education system, largely decentralised in Britain and the United States, is regarded as so vital to national unity that it must be run strictly from the centre - one Third Republic Minister of Education is said to have boasted that, looking at his watch, he could tell which page of which book schoolchildren of a given age were reading all over France. Financial tutelles are also important. Often, it was necessary to gain the approval of the external services for the technical specification of works before the necessary subsidy would be forthcoming from Paris. While in principle both communes and départements possessed 'general competence', being able to undertake any activity in the interests of their populations which was not against the law, in practice this was hedged about with administrative and legal restrictions. Further, their resources were largely pre-empted by their statutory obligations. So the budgets of the départements were largely taken up with social security and highways expenditure, those of the communes on public works programmes encouraged and subsidised by central government.

The prefect, as the representative of the state in each département is a key figure in the jacobin vision. His role was to manage the executive affairs of the département, to co-ordinate the activities of the field services of the various ministries and to oversee the activities of the communes in the interests of national unity. To emphasise his dignity as the representative of the all-powerful state, he lived in some magnificence in the most prominent building in the town and entertained in style. In the past, the prefect acted as the main source of information for the Ministry of the Interior, though with the development of modern communications, this became a largely formal role. He remained, however, the symbol of the one and indivisible republic.

THE BASES OF LOCAL POWER

The jacobin myth is part reality, part idealised vision of what some observers would - or would not - like the system to be. Many observers of French politics in recent years have pointed to contrasting features of the system, not in terms of the formal constitutional provisions but in terms of its political functioning. Wright (1978), claims that the belief in the centralised nature of the French system rests on four factors: the statutory weakness of local councils and the control by prefects and technical field services;

the archaic nature of local government structures; the unrepresentative nature of local elites, depriving them of legitimacy; the financial dependence of local councils on Paris. He goes on, like others, to point out that this gives an inadequate picture of reality. Even before 1981, the tutellary power of the prefect had been much reduced, with mnay new acts simply requiring a formal referral and widespread provision for acts to be validated automatically in the absence of prefectoral intervention. Many observers have drawn attention to the mutual dependence of mayor and prefect. Administering a complex modern state on the basis of unbending uniformity would be a practical impossibility. In order to perform his job, the prefect must make concessions to local pressures (Machin, 1977), and Paris understands this part of his role. Indeed, so great is the fear of prefects 'going native' that they and the sub-prefects are moved around every two years or so. Local politicians themselves are often figures of national substance, members of Parliament, national civil servants or even ministers. Twenty-two of the thirty-one ministers in Mitterrand's first government held local office, seventeen as mayors; these included the Prime Minister, Pierre Mauroy, mayor of Lille and the Minister of the Interior, Gaston Defferre, mayor of Marseille. In 1971, as part of Pompidou's efforts to put down local roots for the Gaullists, thirty-six of the former forty-one ministers stood for <u>mairies</u> (Machin, 1977). In these circumstances, the formal hierarchy is reversed for, as Machin (1977, p 164) comments, 'the Mayor who is a member of Parliament often feels little need to co-operate with the Prefect but rather, expects the Prefect to co-operate with him.' The <u>cumul des mandats</u>, the accumulation of offices, can extend through several levels, with a <u>notable</u> holding communal, departmental, regional and national positions. This can give local elites considerable leverage on the political system, gaining exceptions and special treatment for their own areas and producing differentiated policy outcomes in different localities. So, as Grémion (1976, p153) puts it, the universal rules of the centre dissolve in particularisms at the periphery.

Local power is also encouraged by the fragmented nature of the central administration. The vast expansion of state services at all levels has given rise to new centres of bureaucratic power, centred on the functional ministries or the <u>grands corps</u> of administrators who people them. The prefect's control over the field services of the ministries was always more of an aspiration than a reality and local figures are able to exploit this as well as differences in

Paris (Wright, 1978). It is, indeed, this ability to play on the system which is the essence of the role of the <u>notable</u>, described by Grémion (1976, p 167) as 'a man who has the power to act on the state apparatus at certain privileged levels and who, in return, sees his own power reinforced by the privileges which these contacts, so far as they are sanctioned by results, confer.' The very weak functional basis of French local government, indeed, has meant that local elites have had to seek leverage at other levels; while the practical impossibility of governing a complex modern state according to the pure jacobin model has forced central decision-makers to rely on local collaborators. So, according to Ashford, (1982, p 7), 'while we often think of the Franch political system as more centralised than the British, in practice the policy process in France disperses more power to lower levels of government than in Britain.'

In the big cities, local power may even be more evident. Not only is the mayor likely to be an individual of national substance, but the council will have established its own technical services, so allowing it to by-pass both prefect and the external services of the state. For those matters which the city cannot undertake at its own hand, it can go directly to the highest levels in Paris. The rise of the cities, as we note in the next chapter, has yet further undermined the notion of the unified jacobin state. In many <u>départements</u> too, even before the 1982 reforms, the formal position of the prefect was far from reality. <u>Conseils généraux</u> with active and politically-minded presidents would insist on involvement in drawing up the budget and exert the maximum influence over the prefectoral administration. As in the case of the cities, the ability to do this depended to a large extent on the <u>notoriété</u>, the local and national standing, of the president himself. Nièvre, with Mitterrand as its president, was an active and forceful council, which the prefect would treat with discretion. The more traditional, rural and right-wing or non-partisan councils would be more inclined simply to follow the prefectoral lead. So the jacobin vision of uniformity was far from describing the reality of local political life.

THE POLITICAL HONEYCOMB

The view of the French system as one which provides a real measure of local power has, in turn, encountered its critics. These agree in rejecting the jacobin view of the state as a description of reality but insist that power is not 'dispersed to lower levels of government' but concentrated in the hands of the notables. If an individual holding several offices achieves a policy success, it is often impossible to determine which hat he was wearing at the time. Ashford (1981) extols the role of the mayors in defending the interests of their localities. Yet often the mayor qua mayor is powerless; the maire-député is a figure of substance precisely because of his combination of local with national office. Grémion (1976, p266) points out that 'the power of the notables is not community power. Resting on the mobilisation of bureaucratic resources, it is on the contrary a brake on local communities taking charge of their own problems.' So power is a function of the ability of individuals to play on the complex network of territorial government.

Neither the jacobin view nor an unqualified 'local power' model make adequate allowance for this compelxity. Machin (1979) distinguishes four systems of local government, each providing its own form of legitimacy to policy, the prefectoral system, the parliamentary system, local councils and the field services of the central ministries. Power and policy making can only be understood by reference to the interconnections of these systems. Sociologists of French administration (Crozier and Friedburg, 1979; Gremion, 1976; Worms, 1966; Dupuy and Theonig, 1985) have interpreted the system as a series of complex 'games' in which the actors include local and national politicians (often the same people), the prefects and the members of the grands corps and field services. So the localities deal not with the 'centre' but with a complex network in which power relationships are constantly shifting according to the resources which each participant can bring to bear. The prefects, with their immense prestige, seek to make their nominal control over the field services a reality. The field services, in turn, often more deeply rooted locally (their members do not move about like prefects), seek to maximise their own discretion. They act for local councils on a contractual basis, receiving fees for their work, but their technical approval is often required for the necessary subsidies to be forthcoming from Paris. There are notorious rivalries among field services themselves and among the grands corps from which many of them are recruited. This is the case, for example, between the Ponts et Chaussées and the Génie Rural of the Ministry of Agriculture, which have overlapping responsibilities in rural areas. The Trésoriers Payeurs Généraux of the Ministry of Finance play a key role, exercising an effective financial tutelle over local councils, but they may come into conflict with other ministries. The prefects themselves, far from being the stern guardians of the centralist vision, may take the side of the localities and must, in any casem as we have noted, retain good relations with their local <u>notables</u>.

Ashford (1981) would certainly accept the complexity of the system, but sees this as inevitable in the modern welfare state. The advantages of the French system, in his view, is that, because of the entrenched power of local elites, policy must be negotiated across the territorial dimension. The result may be a more incremental style of policy-making than the British but one with more lasting results. Thoenig (1979) gives a much more critical evaluation, seeing the system as a 'honeycomb', with the virtues neither of local decision-making among peers not of unified vertical bureaucratic control. The system serves the interests of those within it, with a mistrust of debate, excessive secrecy, and access to decision-making limited to <u>notables</u>. If local democracy exists, it is concentrated on access to higher levels rather than local decision-making. Indeed, the system encourages the passing of decisions to higher levels to avoid personal responsibility (Thoenig, 1979). This is a common theme in the writings of the Crozier school as well as of countless folk-tales. A well-known if apocryphal story tells of the mayor faced with a choice between a correct but unpopular decision and an incorrect but popular one. He took the popular decision, quietly asked the prefect to veto it, then publicly attacked the prefect for infringing local democracy.

The consequences of this system, according to its critics, have been a stultifying of local democracy and increasing albeit ineffective, centralisation as issues drift to Paris for resolution. The result is, in the words of the nineteenth century radical Lamenais, 'apoplexy at the centre and paralysis at the periphery' (Hayward, 1983, p 24). In the struggle for influence, the greater <u>notables</u> - mayors of big cities and leading national politicians - win out over the lesser, further eroding the fiction of jacobin uniformity. The power of the <u>notables</u> is self-reinforcing, sanctioned by results, giving most local politicians a long life expectancy (Machin, 1977). It is true that the <u>notable</u> system is beginning to give way, particularly in the big cities, to a more party-political style and this is a development which, as we shall see, the Socialists are pledged to encourage. This has served further to differentiate the politics of the cities from those of the rural areas, entrenching divisions between town and country and, rather than sweeping away the <u>notable</u> system, has generated new forms of it. <u>Notable</u> power, in its various forms, is strongly entrenched, including within the <u>Parti Socialiste</u> itself.

Nowhere more clearly can the resilience of the system be seen than in its capacity to resist reform. Alone among Western European countries, France has not experienced a major reorganisation of its local government system in the post-war period and this despite almost universal agreement that the many small communes are hopelessly inefficient and can never hope to enjoy real independence, and that departmental boundaries, laid down in Napoleonic times to allow a man to ride from any point in the <u>département</u> to its capital and back in one day, need to be revised. The regional level, widely seen as more appropriate to the needs of modern planning and administration, was for a long time rejected and, generally, over large parts of the country, the local political system was out of line with contemporary social, economic and political realities. The main problem is that the local political elites themselves, while rhetorically declaiming against the centralised state, are its main beneficiaries, once they have learnt to play the appropriate games (Grémion, 1976). They are then able to use their position in national politics, notably in the National Assembly and, above all, the Senate, which is elected from local councils, effectively to block change.

THE SOCIETE BLOQUEE

We have examined three models of the French centralised state, one, the largely ideological jacobin model, stressing centralisation and uniformity; one stressing the existence of local power whatever the formal provisions; and one stressing the concentration of power in a political/administrative class which is able to operate at all levels of the system, hiding its operations behind the rhetoric of the first two models. It is the criticians of this third school which have been most influential in recent debates on French centralisation. As we have noted, these do not claim that France is a jacobin state; far from it. They recognise the complexities of power and its dispersal in intricate networks as well as the diversity of policy outcomes throughout the country. The system is nevertheless seen as centralised in that decisions do tend to drift through the networks to Paris - but, in contrast to the jacobin model, they are not taken consistently. Individuals can escape responsibility for their own decisions while claiming credit for favourable ones. Local institutions are weakened by the <u>cumul des mandats</u> and together with the opacity of decision-making this weakens local democracy. Citizen participation is discouraged and political choices obscured. Horizontal control and planning of administration at the local and regional level is assured neither by the prefect nor by the councils. At the centre itself, the preoccupation with local matters congests the machine and prevents strategic initiatives unless these are undertaken by special agencies, by-passing the political system - it is pointed out that France was the last major western country to complete its strategic motorway network. The phenomenon of the societe bloquee leads to a stifling of initiative and innovation. It is a heavy list of charges but one which, in various forms, attracted a good deal of support in the France of the 1970s. We shall see in chapter three how the Socialist Party came to abandon its jacobin legacy and make decentralisation one of its priorities for government. They were by no means the first, though, to try and tackle France's centralist tradiation and we must first examine the history of previous attempts. We must also, to avoid becoming ourselves prisoners of the jacobin myth, note the substantial changes, institutional and behavioural which the system has experienced over the years and the changing economic and urban political environment in which it operates.

2 Before Mitterrand

The declared intention of the French Socialist administration, presided over by Mitterrand, was 'to give the state back to the people' (Ardagh, 1982). It is doubtful whether the people of France ever possessed the state, but with this initial proviso, the underlying message is clear enough. Elected to office in 1981, the French Socialists proposed to construct a 'new citizenship' incorporating new structures of participation, expression, communication and democracy. Central to this aspiration is the promise of change and decentralisation, manifested in the lengthy, controversial programme of legislation ushered in by Gaston Defferre. The programme is complex and will be discussed mainly in Chapter Five. It needs to be seen against a backcloth of increasing state centralisation - often under the guise of reform, planning or modernisation - and years of dispute over the relationship of the French state to its citizens. The purpose of this chapter, therefore, is to situate the prospect of change and decentralisation within its historical context.

THE CENTRALISED STATE

According to Mitterrand, strong centralisation was necessary for 'the making of France' but, equally, decentralisation was now essential if France was not be 'unmade'. There are, as we have noted, numerous 'myths' surrounding the nature and intensity of French centralisation and nuances, subtleties and complexities which soften the commonly paraded view of France as the archetypal centralised state in Western Europe. However, the centralised nature of the French administrative apparatus is not seriously in doubt. More often than not, this situation is attributed to the French Revolution, which unified French administrative structures, and to Napoleon Bonaparte, who resorted to military inspired thinking to reinforce this process. Of course, the path towards a centralised French state began somewhat earlier, reaching its pre-Revolutionary apotheosis in the <u>ancien régime</u>. The latter represented an aggregation of provinces built up slowly, unevenly since the tenth century. The creation of the French nation-state was accelerated in the seventeenth and eighteenth centuries, notably via Bourbon monarchs (Louis XIII and Louis XIV) and their ministers - Richelieu, Mazarin and Colbert.

The Revolution, of course, swept away the <u>ancien regime</u> whilst consummating the unfinished process of nation-state building begun by its predecessors. The Jacobins nurtured decentralist aspirations but, faced with provincial counter-revolution, created a unitary state - 'the one and indivisible Republic', based on the ninety <u>départements</u> to preserve the gains of the Revolution and so paved the way for Napoleon's militaristic centralisation of the French state. The Convention had created <u>représentants en mission</u> to supervise the breaking of the counter-revolution. Seven years later (1800), by the <u>loi du 28 pluviose</u> (Year VIII), Napoleon created the <u>préfets</u> (prefects) to act as his 'little emperors' in the <u>départements</u>. The prefects, whose predecessors were the <u>intendants</u> of the <u>ancien régime</u>, functioned as central state representatives responsible for such duties as law and order, information and executive power in the <u>départements</u>. The prefect was charged by Napoleon with a political and administrative role. The institution represented the authoritarian and centralised face of French public admnistration. Indeed, Napoleon had been impressed with the Roman origins of the prefectoral role, so the Emperor accomplished the centralist objectives of his monarchical predecessors, even relying heavily upon the personnel of the <u>ancien régime</u> to fill his administration and prefectures.

The institutions of these years - prefects and <u>départements</u> - served France for the next one hundred and eighty years. In fact, the Napoleonic period also established a durable pattern of power distribution for, according to Bourjol (1969): 'The <u>département</u> was no longer a tier of decentralised decision-making, it became a tier of authority (commandement), a vehicle for the implementation of decisions from the centre (pouvoir central.)' In the post-Napoleonic era, there were some developments in the field of public administrative reform but none of these represented significant decentralisation. For instance, the July Monarchy yielded minor concessions to administrative decentralisation and deconcentration. In particular, legislation passed between 1831 and 1838 allowed for greater recognition of the département's status as a basic administrative unit of the French state (a collectivité territoriale) and the election (albeit by limited suffrage) of the departmental general council (conseil général). The prefectoral institution survived these years and, except for a brief three-month period in 1848 when they became re-entitled commissaires de la République, were charged by the Second Republic's Minister of the Interior (Ledru Rollin) with the defence of the Revolution and Republic 'by their acts, their words, their teaching' (Bernard, 1983).

The Second Empire (1852-1870) signalled the return of Bonapartist authoritarian rule and although decentralisation was often discussed it was little practised. However, there were some decrees which amounted to deconcentration whilst towards the end of the Second Empire, there was some liberalisation for general (departemental) and municipal (communal) councils. By 1870, debate had centred on the respective merits of conferring primacy within the departmental framework to the préfet or the conseil général. This discussion recurred in the Third, Fourth and Fifth Republics and raised its head again with the Defferre proposals, discussed below (Chapter 5).

The Third Republic, of course, retained the prefect as the repository of state power and also made the prefect executive of the collectivité départemental. Despite this perseverance with prefects, who were empowered to exercise a tutelle (supervision) over local authorities (collectivités locales), they should not be seen as all-powerful beings. As Machin (1977) explains, by the end of Napoleon III's rule, the myth of the omnipotent prefect 'bore little resemblance to reality'. If this was the case in 1870, it was increasingly so during the Third Republic. Certainly, the prefect of the Third Republic emerged more as an administrative than a political figure. In any case, the prefect's powers were limited by the activities of députés, central government, local administrative and political forces. In 1871, for instance, the Constitution provided for a commission départementale elected by the conseil général and able to control and call to account the prefect. This provision led Adolphe Thiers to protest against 'these departmental commissions, like so many pains in the backside of my prefects' (Bernard, 1983). Nevertheless, this was the price to be paid to the understandable reaction against the personalised authoritarianism of the Second Empire.

As to collectivités territoriales (départements and communes), Machin maintains that, in the early years of the Third Republic, the prefectoral control of former times was replaced by a mild form of prefectoral supervision. Moreover, the celebrated laws of 1871 (on départements) and 1884 (on communes) guaranteed direct, universal suffrage - the local, democratic, elective bases of the Third Republic. In addition, in 1875, the Senate was elected by representatives of the collectivités locales making it, in Gambetta's words, the grand council of French Communes and a veritable preserve of the notables. In 1871 law (la grand charte départemental) recognised the conseil général as a valid deliberative body equipped with powers to vote the departmental budget. Similarly, the 1884 law empowered municipal councils to deliberate 'local affairs'. These two historic laws recognised the 'moral personality' (personnalité morale) of départements and communes as collectivités territoriales and, henceforth, were help up as shining examples of decentralisation.

In practice, the important legislative gains of the young Third Republic were diluted over time by the growth of the state's financial, administrative and technical tutelle (Mény, 1974). Increasingly, Paris intervened to expropriate various services and duties or impose diverse planning, development or technocratic imperatives. Small communes, in particular, proved susceptible to encroaching central state demands and expenses, thereby undermining the formal equality of communes embodied in the 1884 law. Furthermore, the powers of economic intervention of collectivités locales were monitored strictly by the state. In this respect, the prefect's tutelle could be irksome. Nevertheless, the départements and communes of the Third Republic enjoyed at least nominal autonomous status as collectivités territoriales. The same could not be claimed for the region to which we now turn.

THE REGION

The Mitterrand 'experiment' intends to redress the history of French regional policy by placing the region on an equal footing with the départements and communes (Chapter 5). This promotion for the region consumates the rapprochement between the French Left and the regional lobby during the Fifth Republic. This entente cordiale, principally a child of the 1970s, has not always been possible. Traditionally, the Left has equated regionalism with provincial counter-revolution and right-wing reaction against Republican gains. Neither The French Revolution, nor for that matter Bonapartism, had much sympathy for provinces or regions as intermediary levels of the unitary nation state. The Third Republic followed suit.

The idea of regionalism became associated closely with Maurras, Barres, Renan and the Vichy period. According to Gourevitch (1977), the Right developed 'a nostalgia for a supposed decentralised, free, flourishing, provincial past'. In practice, this meant rule by local elites. Nevertheless, the reluctance to elevate the status of the region did not signify a lack of interest in the idea of regionalism. In fact, Bourjol notes forty-eight texts on regionalism submitted to the National Assembly between 1871-1940, 'a very rich period in the sphere of regaionalism'.

These proposals differed widely in context but common to most of them was the desire to create areal units (about eighteen to twenty-five) to supplement and/or replace départements, which were criticised as outmoded, geographical, administrative units. Moreover, the new regional unit would be designated a collectivité territoriale. As we shall see (in Chapter 6), these potential developments were stillborn initiatives until the 1980s, when Corsica supposedly led the way for the rest of France. In the Third Republic, some of the projects condemned départements and/or préfets, proposing to surpass, abandon or amalgamate (where appropriate) the artificial structures of the Revolutionary/Napoleonic era. Only a handful of timid regionalist proposals actually reached the statute books after 1871 aand, significantly, these owed their 'success' to right-wing majorities, thereby confirming left-wing mistrust of regional demands. Essentially, the Third Republic rejected a fully fledged regionalism and enhanced status for the region. However, we see emerge or consolidated in this period, many of the ideas and institutions - for instance, regional prefects and the nomination of interest groups to regional councils - which have to wait until the fall of the regime in 1940. Further, the arguments of the day (which will be developed throughout this chapter and volume) can be follwed through to the postwar setting: the break-up of France? the potential over-politicisation of French public administration? too many layers of government and administration? directly or indirectly elected or appointed regional bodies? an adequate provision of local finances? Critics of regionalist policies saw in them a threat to the French nation-state. Not until the postwar times did it become obvious that regional 'concessions', in the form of decentralisation, could serve as the alibi of state centralism rather than its negation. In short, the elasticity of the regional idea was underestimated.

Understandably, the region versus département argument underlay many of the proposals after 1970. For example, the Hennessy Report (1915) expounded the impossibility of balancing equitably the region against the département and came down in favour of the region: 'the region must, in the thinking of the legislators of 1915, meet the needs for which the men of 1789 set up the département' (Bourjol, 1969, p 181). In the 1980s, the Mitterrand government hoped to bypass this dichotomy between region and départment by counter-balancing all levels of administration. Of course, this broad principle of a place for everything (and everything in its place) contradicts most of the thinking on the Left during the century following the birth of the Third Republic. A major influence on the Left's perspective was the Vichy experience, to which we now turn.

VICHY AND AFTER

In the 1930s, regionalism came to be discussed (for example, in the Doumergue project of 1934) increasingly within the context of an authoritarian state. This prospect was realised with l'état français of 1940-1944. Inevitably, the experience of Vichy regionalism and the behaviour of some regaionalist movements under the German-Italian Occupation coloured the lenses of the postwar French Left. Under Marshall Petain's Vichy administration, debate on regionalism as an economic or administrative option was supplanted by the practice of regionalism as a technique de commandement within a strong, centralised state (Bourjol, 1969): prefects were reinforced and politicised; regional prefects were created to carry out certain administrative and political tasks - including police duties, economic affairs and food rationing; conseils généraux lost power to the prefects and various delegations of a departmental or regional nature were set up to liaise directly with and reinforce central government. In theory, the region enjoyed an elevated status but, in practice, this meant little since the Vichy region functioned as a mere appendage of central government.

After the Vichy phase, there followed a brief (June 1944 - March 1946) period of regional, political deconcentration as the French Provisional Government and the Fourth Republic persisted with inherited Vichy structures in the form of commissaires de la République. These were de facto regional prefects located in the main urban centres and wielding considerable powers in matters such as law and order, the economy and control of various services. Bourjol (1969) draws historic parallels with the aforementioned représentants on mission of 1792, despatched by the Revolution to the unruly provinces. In both times of civil unrest, exceptional structures were considered expedient. However, the memory of imposed Vichy regionalism was distasteful for the politicians of the nascent Fourth Republic and, consequently, the commissaires de la République were abolished as soon as the haggling over the new Constitution terminated.

Arguably, at this point, the baby was thrown out with the bathwater. The Constitution of the Fourth Republic provided little comfort to regionalists and signalled the revival of prefects, albeit under the official, constitutional guise of délégués du gouvernement. Départements and communes were confirmed as the basic units of the French state (collectivités territoriales) whilst a change in the Constitution was required if regions were to acquire this status. Alternatively, regions could be formed as subordinates or groupings of départements.

There were obvious parallels with the birth of the Third Republic. Anxious to control power following a dose of strong government, the Fourth Republic's constitution makers modified the role of the prefect who became co-ordinator, not the head, of state services. Further, the weaker concept of controle administratif replaced the former tutelle. In principle, there was scope for the extension of the powers of conseils généraux but this did not materialise really - due to local default and to the recentralising tendencies of the state. In contrast, the délégué du gouvernement (or prefect) was able to maintain and consolidate his position in the ailing Fourth Republic. Nevertheless, he shared with locally elected representatives a common enemy - the accelerated growth of central and field services of the state and the effective erosion of power via 'advice' from planners, technocrats, experts and agents of the state.

In postwar France, a familiar pattern soon emerged as the centralised state proved reluctant to spread its load more extensively. In 1948, this trend underlay the regaional experiment of IGAME's (inspecteurs généraux de l'administration en missiom extraordinaire). Bourjol dubbed the new form of regionalism as 'regionalism without the region' since the institution (IGAME) was created before the actual formation of regional boundaries. As with Vichy, regionalism was restored as a means of command. Effectively, each IGAME functioned as a super-prefect in charge of an IGAMIE, which was a grouping of from six to thirteen départements - making nine IGAMIEs in all. The main duties of the IGAMEs concerned aspects such a police, security, advice and information - but also economic regionalism and co-ordination. Therefore, the IGAMIE represented a sort of military-cum-civil grande région, created not for the cause of regionalism and decentralisation, but rather for the protection of the Republican state. Of course, 1947 was a time of industrial, economic and civil unrest within France and a tense international climate without. These circumstances led to the IGAME experiment and provoked the title of 'cold war regionalism' (Bourjol, 1969). Initially, the new inspectors (IGAMEs) were based in Paris and might be seen as a threat to prefectoral power. Before long, however, thay became regionalised under the auspices of the main prefect of the 'region' and, in any case, enjoyed support from the prefectoral corps, which approved of the welcome boost for law and order. According to Meny (1974), the IGAMEs inaugurated the postwar era of regional deconcentration (as opposed to genuine decentralisation), in which regionalism was imposed by forces actually hostile to its principle.

Without doubt, the trend towards a co-ordinated, concerted economy was strengthened by the IGAME experiment, which lasted until the reforms of 1964 (see below). Further regional initiatives confirmed this direction. For instance, a major concern of the Fourth Republic was to decongest Paris and alleviate the problems of the provinces, the infamous desert français (Gravier, 1972). Thus, in 1950, aménagement du territoire (regional planning) was adopted by the Ministry of Reconstruction and Urbanism. This paved the way, in 1954, for Edgar Faure's introduction of Fonds nationale d'aménagement du territoire (FNAT), which included a mixed bag of incentives, loans and aids to ailing communes and départements, disincentives for developments in Paris and fiscal carrots for industrialists willing to incest in crisis regaions. The next stage was Pierre Mendès-France's adoption of comités d'expansion économique

(1954) - a diverse body of regional, departmental and socio-professional elite bodies amounting to a veritable 'industrial kaleidescope' (Meny, 1974). Subsequently, in 1955, twenty-two economic planning regions were mapped out and programmes d'action régionale (PARs) created to promote soci-economic expansion in areas of high unemployment and underdevelopment. Unfortunately, the PARs were harmonised badly with their successors, the plans d'aménagement régional (1957) - launched also to offset regional decline. Other bodies, too, such as the sociétés de dévelopement régional (SDRs), set up in 1953 to encourage private savings in poor regions, shared in the task of regional redress.

Discussion of the above (and other) innovations is beyond the scope of this chapter but, significantly, Gravier (1972) complains of 'catalogues of good intentions' and Meny (1974) interprets the reforms as devoid of any political, theoretical or philosophical meaning. Therefore, the end result was 'institutional anarchy' as numerous criss-crossing layers and responsibilities co-existed unevenly. To be fair, there were some achivements in the area of industrial decentralisation but, overall, the advocates of change and decentralisation remain unsated. The Fourth Republic's weak political system retreated from bold decisions in this respect. There was no significant increase in powers for general or local councillors whilst any regionalist measures were 'empirical, fragmentary and évolutif' representing instruments of state adaptation to postwar realities and planning priorities (Meny, 1974). In short, the Fourth Republic rejected decentralisation and settled for economic neo-regionalism and deconcentration within the parameters of a dirigiste economy.

THE GAULLIST FIFTH REPUBLIC (1958-1969)

The critical transition from the Fourth to the Fifth Republic entailed a decisive change in constitutional-political structures. The resort to Gaullist personalised, authoritarian rule included the rewriting of the Constitution, the provision of emergency powers, a change in the voting system from proportional representation to majority voting (scrutin d'arrondissement) in order to encourage stable governments, limitation on the hitherto powerful National Assembly and (in 1962) the direct election of the president by universal suffrage. These measures, of course, amounted to a telling reversal of the Fourth Republic's record. To what extent did the new regime signal change in the practice of decentralisation?

Dr Gaulle was not interested primarily in decentralisation until late in the day. However, the Gaullist decade was by no means barren of measures in the domain and, ultimately, proposals for reform in this sphere led to the General's political exit (Hayward, 1969b). In the early years of the Fifth Republic, prior to the 1964 reforms, the regime pursued a policy of administrative deconcentration and state regionalisation. In fact, Bourjol (1969) defines the policy of the Fifth Republic as a second phase of neo-regionalism synthesising Vichy corporatist neo-regionalism and the economic regionalism of the Fourth Republic.

The government recognised the need to harmonise and rationalise administrative units and regional policy. In 1959, therefore, this involved a better definition of France's twenty-one regions in order to promote socio-economic development and aménagement du territoire. Moreover, whilst the prefects maintained their status, special préfets co-ordonnateurs were placed at the head of each region. Their duty was to look after individual départements but also to convene interdepartmental conferences drawing together interested parties. In 1964, the préfet co-ordonnateur became officially the regional prefect (préfet de région) with the responsibility for promoting economic development and aménagement du territoire, upholding the policies of the government and easing prefectoral communication through conferences administratifs régionales (CARs). The CARs acted as a regional extension of central government, drawing together departmental prefects and treasurers in each region and possessing an important role in allocating moneys to départements and communes.

To assist the préfet de région, the mission régionale served as a specialist, back-up think tank emanating from the various, centralised ministries. In addition, the regional prefect would preside over the newly created CODERs (commissions de développement économique régional), which replaced the former comités d'expansion. The CODERs combined locally elected representatives (25 percent), government appointees (25 percent) and socio-professional nominees (50 percent) and may be seen, optimistically, as a step towards regional recognition. However, their role was purely consultative. Budgetary powers were lacking and, in practice, the CODERs were subject to prefectoral manipulation and notables' opposition, sabotage or eventual colonisation. The notables opposed institutions which disturbed their traditional power bases and the prefectoral corps resented the notion of hierarchy implicit in the institution of regional prefects. In practice, the 1964 reforms consolidated the initial Gaullist phase of deconcentration to the prefect's benefit but again, it is worth underlining that the prefect's real powers should into be exaggerated. He had to work with local interests, and more than ever, faced the growth of specialist 'advice'.

One limitation on the prefect was provided by the growth of city councils and their influential mayors. Machin (1979) points to a massive expansion of local government staff: 'Instead of replying on field services the city councils have built up their own teams of economists, though the abundance of communes (see below) left the prefect in a powerful role as arbitrator, he needed the support of powerful notables - not least, those benefiting from the multiple office holding tradition (the cumul des mandats). Therefore, the prefect should not be seen as an enemy of the local representatives he had to work with. On the contrary, as Worms (1966) and Gremion (1976) illustrate, there existed a relationship of complicity, solidarity, reciprocal dependence and the status quo against would-be power usurpers such as the specialist technical services emanating from government ministries or the so-called forces vives, new social layers prepared to 'accept their responsibilities' in the political arena. To some extent, regional demands could be interpreted as the desire of the forces vives to acquire position of influence in French administration.

With the 1964 decrees, many critics also opposed the creation of a further level of administration (the region) for, arguably, 'there had been a move from departmental deconcentration to regionalisation' (Bernard, 1983). Retrospectively, the CODERs may be seen as stepping stones to the idea of stronger regional entities with real powers, directly elected assemblies and executive authority. At least, as Mény (1974) contends, the region was placed on the agenda as a suitable unit of deconcentration and planning. The decentralisation proposals of Defferre bear little comparison with the 1964 reforms but do build upon the weakness and defects of the latter. The main problems with the 1964 decrees were summed up effectively by Wright (1979): 'They remained unrepentantly rooted in central state supremacy: there was no independent regional budget, there was no regional services and there was no attempt to create a collective regional conscience'.

Similar criticisms could be applied to economic planning in the Fifth Republic. Planning is dealt with elsewhere

(Chapter 7). Suffice it to say here that the early years of the Fifth Republic witnessed the sponsorship of <u>plans régionaux de développement économique et social d'aménagement du territoire</u> (that is, regional planning over longer timespans that five years), the regionalisation of the national plan via <u>tranches regionales</u> and the creation of the DATAR (the <u>Delegation à l'Aménagement du Territoire et à l'Action Régionale</u>), which consummated the Fourth Republic's moves in this directiotn. Born in 1963, the DATAR acted as an organ of regional co-ordination and impulsion and was answerable directly to the premier, thereby indicating its centralised basis.

The net effect of the panoply of Gaullist measures and reforms was to turn regionalism and decentralisation into the instruments of the Gaullist state. Certainly, statism was given a regional face - but mainly to suit Paris rather than to meet the lobby for effective decentralisation. A decade of Gaullist rule yielded some administrative reforms and recognition of a regional dimension but failed to dampen cries for a more fundamental retructuring. In 1968, prior to the May 'events', de Gaulle dropped his regional bombshell. In March (at Lyon), the General explained: 'The centuries-old centralising effort required to achieve and sustain (France's) unity is no longer necessary. On the contrary, it is regional development that will provide the motive force of its future economic power'. (Hayward, 1983). A fortnight later, de Gaulle's speech was backed up strongly by Olivier Guichard (the first head of the DATAR and a leding Gaullist 'baron'). This Gaullist thrust was open to interpretation. Did it represent simply a concession to liberal Gaullists? Was it a response to the regional lobby and policy critics? Could it be seen as a genuine attempt to decentralise power? Whatever the verdict, it offered hope to regionalists - not least since it forced the French Left to examine belatedly its conscience on this matter.

The Lyon speech was quickly overtaken by the May 1986 crisis, construed widely as a challenge to (Gaullist) state centralisation and authoritarianism. At the height of the May unrest, de Gaulle looked to reinforce prefectoral, state power, but with the storm 'over' and a huge June electoral success under his belt, turned once again to regaional reform. Regrettably, the issue of regional reform and decentralisation became complicated with the issue of 'with or without de Gaulle' and the controversial resort to the referendum and the concomitant replacement of the Senate (Hayward, 1969b). The Gaullist referendum of 1969 thrust the debate about regionalism and decentralisation to the forefront of French politics. The proposal to make the region a collectivité territoriale, with jurisdiction over economic, social, cultural matters and aménagement du territoire, was highly controversial. The region, hitherto seen as a collection of départements and communes, would benefit from a transfer of state power enabling it to control areas such as construction and public investment (including roads, bridges, schools, hospitals and airports). The new collectivité territoriale was not to enjoy its own fiscal resources but, by way of compensation, state taxes would be transferred to the region in accordance with the responsibilities granted.

The Gaullist proposals for reform of regional policy were not unpopular. Opinion polls confirmed a majority of the population in favour of decentralisation. The problem lay in the unfortunate association with de Gaulle's political survival and the reform of the Senate, France's second chamber. The reform of the Senate paralleled the regional proposals in that the new second chamber would combine senators (elected by the limited suffrage of municipal and general councillors) with the appointed state consultative body, the Economic and Social Council, which represented socio-professional organisations. The new, revamped Senate was to become a regional council body rather than retain its status as a grand conseil des communes de France. De Gaulle's proposals envisaged the new body as a mere consultative assembly, flanked by mini-senates in the regions. All in all, the propositions were complex, contestable and unnecessarily confused. For example, even the region - despite gaining nominal recognition - was not granted full competences, a legislative role or a directly elected assembly, therefore leaving it in practice in a constitutional no-man's land somewhere between a collectivité territoriale and the less elevated établissement public (Mény, 1974). Moreoever, according to one observer (Bernard, 1983), 'The referendum campaign was launched in a climate of political passion, legal dispute and national drama.' Consequently, the 1969 (April) Referendum resulted in a marginal defeat for de Gaulle - 53.17 percent voted no to his proposals, on an 80 percent electoral turnout.

L'APRES-GAULLISME

After the Gaullist reversal of 1969, President Pompidou turned his back on regional reform - against the liberal sympathies of some of his supporters, including Jacques Chaban-Delmas, his Gaullist premier. Pompidou was unwilling to risk losing the support of forces disturbed by de Gaulle's regional policy, yet still receptive to a Gaullist-led majorité. These elements including Centrist notables, the prefects, Gaullist Jacobins (Michel Debre, Jacques Chirac) and anti-regionalists. In practice, Pompidou could agree with the claim that 'the majority of politicians and civil servants have no desire to tackle this sensitive problem, and their inaction is widely approved by conservative local notables and apathetic electors'. Therefore, instead of ambitious decentralisation and regionalism, Pompidou opted for technical deconcentration. In 1970, this included public investment in the regions, départements and communes as regional and departmental prefects acquired control over funds (dotations globales) delegated by the state and hitherto controlled in Paris. Pompidou supported some degree of decentralisation to the département and commune. Thus, from 1970, the conseils généraux of the départements were to be consulted in planning and construction programmes, the mayors' powers were boosted and the municipal councils relieved of the tutelle. In 1971, however, the attempt to fuse and regroup France's communes (via the Marcellin law) ran into much opposition from local representatives. Of course, reform of municipal boundaries is a particularly sensitive issue since many (about 80 percent) of the French communes dated back nine hundred years or so. In contrast, the region lacked similar pedigree unless it could be equated with historic provinces.

Pompidou continued to view the region as 'a harmonious expression of the départements which together comprise it'. The region was not to be given the status of a collectivité territoriale. Instead, in 1972, it was designated as établissement public, whose function was to contribute towards the economic and social development of the region. Clearly, this left the region in an inferior position vis-a-vis collectivités territoriales and signalled a retreat from the 1969 proposals. This situation was further accentuated by the composition of the new regional council (conseil régional), comprising members of the national legislature (50 percent) and members elected by general and municipal councils (50 percent). No provision was made for directly elected regional councils and, moreover, the conseil régional lacked basic stability since its make-up was subject to electoral turnover. In addition, there was to be a new economic and social committee drawn from regional socio-professional groups and mirroring the national Economic and Social Council. Last but not least, the prefect remained the chief executive of the region.

Overall, the measures adopted under Pompidou's presidency left the region as a level of co-ordination and concertation - but not decentralisation. Unsurprisingly, Wright and Machin (1975) sum up Pompidou's 1972 reforms as 'a case of disguised centralisation'.

The election of Valery Giscard d'Estaing to the presidency in 1974 was viewed by diverse advocates of regionalism and decentralisation as a cue for change. For one thing, Giscard came to the presidency with a reputation for supporting greater decentralisation. Also, he had criticised his predecessors' timidity in this respect. Certainly, there was room for decentralisation in Giscard's liberal blueprint for France and certain ministerial appointments enhanced this prospect. A potential coup was the recruitment of arch-regionalist Jean-Jacques Servan-Schreiber as Minister of Reforms. Nevertheless, in the field of local and regional reform, Giscard demonstrated remarkable continuity with his predecessor. Like Pompidou, Giscard was anxious to placate notables, Gaullists and Centrists, who felt threatened by the possibility of political-administrative reform. His narrow presidential victory rested on a fragile political majority. Moreover, as the Left gained local electoral successes, the Right did not want to see any decentralised powers passing into the 'wrong hands'. The Italian experience of Communist regional advances offered further and sobering food for thought.

In practice, therefore, the Giscardian septennat(presidency) confirmed Gourevitch's observation that the 'demand for change tends to come from those out of power. Once in power, they find centralisation useful' (Gourevitch, 1977). The ministerial role of Servan-Schreiber lasted a few days only and the regime declined to pass along the path of regional reform and decentralisation. In 1975 at Dijon, Giscard informed regional representatives: 'The role of the region is not to administer nor manage itself, nor to substitute its intervention au pouvoir de decision des collectivités locales, which must be developed and reinforced. Its role is to assure, at an appropriate level, the co-ordination of our economic development' (Bernard, 1983). This limited regional perspective was shared by Gaullist premier, Jacques Chirac, who opposed stronger regional entities. In 1976, Chirac's message to a gathering of the Presidents of Regional Councils was that regions were 'artificial structures' and as such would be denied the status of collectivités territoriales. As for those other once-artificial structures, the prefects and départements, these increased in status under Giscard. In contrast, the commune – France's most durable unit of administration – underwent mixed fortunes and deserves some comment below.

The French Revolution had abandoned duchies, provinces and counties for uniform departmental structures but retained the communes as basic units of administration. By the Gaullist era, these somewhat dated entities were largely unable to withstand depopulation, désertification, modernity, and rapid urbanisation. Many of the thirty-seven thousand communes were unviable as efficient administrative units and lacked the resources to provide an equitable range of services. The Gaullist regime had undertaken the task of overhauling and rationalising French politico-administration structures in order to assist socio-economic change and modernisation. For example, this involved leaning on urban and surburban communes to merge into metropolitan bodies (communautés urbaines). However, reform of municipal boundaries is a particularly touchy issue in France since many of the communes (80%) date back nine hundred years or so. 'Late' Gaullism was more sensitive to locak feeling and, as Ashford (1982) explains, it was left to Pompidou to restore the primacy of the commune as the basic unit of local government. Pompidou contined to encourage communal fusions and mergers (notably via the Marcellin law of 1971), in the form of Sivoms (Syndicats intermunicipals a vocation multiple) – but the emphasis was on voluntarism, plus checks and safeguards. Some moderate successes were in evidence though overall, the communes proved resistant to any high-handed attempts to alter their historic status.

Pompidou proceeded cautiously with the modernisation of French local government and his premature death prevented significant changes under his presidency (Ashford, 1982). Nevertheless, the Peyrefitte Commission on Decentralisation began to question the local government system and urged a better definition of 'local affairs' – significantly, a cause adopted later by Mitterrand and Defferre. Under Giscard, background work continued in the form of the Commission on the Development of Local Responsibilities, which produced the much discussed Guichard (1976) Report, Vivre Ensemble. Guichard made various proposals and criticisms calling for a modernisation of French local administration. In particular, Vivre Ensemble supported fiscal and financial powers for communes; abolition of the tutelle; stronger conseils généraux; a comprehensive, local civil service; and the voluntary creation of communal federations. However, Guichard's intended révolution tranquille was unpalatable, untimely and unacceptable for the ruling, political majorité. The Left, too, attacked Guichard on numerous points - bureaucratisation of local affairs; inadequate, local financial resources; failure to limit the technical tutelle of ministerial services; derogation of communal specificity; and a suspect redistribution of power. Ironically, some of Vivre Ensemble's proposals enjoyed favour after 1981 although Guichard (Le Figaro, 1.12.1982) actually attacked the timidity of Defferre's framework law. Instead of adopting Vivre Ensemble, Giscard and his ministers launched various initiatives, between 1977 and 1981, to consult with and reform local government. They amounted to little, faced with considerable opposition from friends and foes alike (Ashford, 1982). The Senate, in particular, continued to defend the status quo, proferring hundreds of amendments to government legislation. In May 1981, therefore, the programme of administrative reform remained quite incomplete.

The responsibility for reform now fell to the Socialists, who approached matters of decentralisation and regionalisation with a new vigor and a decade's sympathetic reflection behind them. The Socialists' decentralisation programme determines the content of the rest of this book but, at this point, it will be instructive to pinpoint some of the main difficulties bound up with this controversial policy area. The Socialists' blueprint needs to be seen against historic and recent dispute over the make-up of the French state and bearing in mind Ashford's (1982) proviso that 'few reforms send more tremors through the fabric of French political institutions than proposals to restructurel local, departmental...., or regional government'.

Conclusion: the road to 1981

Certain themes predominate in the pre-1981 pattern of French public administration. There has been little, significant, real transfer of powers to départements, regions and communes. On the contrary, the state has managed to retain and strengthen its role via various devices and practices. The Fifth Republic illustrates this process and, according to Birnbaum (1977):

> Although during the Fifth Republic the state has often entertained discussions of decentralisation, the new forms of action by the administration, including both the techniques used and its tutelage over the new regional public institutions, along with the lack of

financial aid to the peripheral administration not to mention the public institutions or even the local collectivities which cannot increase their resources or revenue, have led to the further strengthening of tendencies which encourage segmentation in the state apparatus in France.

This analysis is supported by numerous students of French politics and administration, notably Gourevitch (1977) who explains that 'despite the clamour for change, despite the endless complaints about centralisation, despite the popularity of such general schemes of regionalisation, the pressures for reform have been channelled into safe waters.' In particular, besides the unwillingness to decentralise (as opposed to deconcentrate) powers, we can observe a persistent refusal of the region as a new basic level of administration. In part, this reflects the tenacity of existing power holders (such as prefects, notables, state representatives) in holding on to their spoils. In part, also, it reflects a lukewarm attitude on the part of those in high office towards decentralisation and regionalism. However, this did not preclude a recognition of sorts that the département was too small as a unit of administration for the twentieth, even the late nineteenth, century or that the modern state was an increasingly top-heavy and bureaucratic apparatus. Consequently, the region has been adopted by various governments as a unit of economic planning, administrative deconcentration and an appendage of centralised decision-making. Decentralisation for the region has been resisted en haut - especially demands for regional status as a full collectivité territoriale incorporating directly elected regional assemblies and executive power.

The lobby for decentralisation and regionalism also suffers by virtue of its divisions. Technocrats, planners, politicians, autonomists, separatists, ecologists, autogestionnaires and others swelled the ranks of the regional cause - but for different reasons and with different purposes in mind. As we have noted, the French Left espoused the cause belatedly and unevenly on account of the nostalgia for Republican forms and the discouraging association of regionalism with right-wing, counter-revolutionary reaction. The rapprochement of the late 1960s and the 1970s placed some regionalists in the left-wing camp and there was correlation between the Left's adoption of regionalist and decentralist themes and the long march to power via the stepping stones of municipal and departmental elections. A significant feature of electoral behaviour in the 1970s was the Left's success in previously right-wing strongholds, for example, the West of France. Victories in 1981 consummated the long haul to high political office and some regionalist and autonomist groups (see Chapter 6 on Corsica) were able to claim their place in the Mitterrand majority of May 10.

Once in office, the Socialists would be expected to look towards their election promises, enshrined in the '110 Propositions' of Francois Mitterrand. Propositions 55-59 (Des contre-pouvoirs organises; un état décentralisé) urged decentralisation as a priority and regionalism of meaningful proportions. These goals were central to the Socialists' projection of 'a new citizenship'. The revolutionaries of 1789-1792 had resorted to centralisation to preserve the rights of citizens (as opposed to subjects). Two centuries later, the French Socialists were equally concerned with rights and citizenship but, in theory at least, via a different, decentralist path.

3 Regionalism and micronationalism in mainland France

REGIONS AND FRANCE

France, in Hayward's (1983, p. 21) words, is a 'unitary state imposed upon a multinational society'. The nation is 'an artefact of the state', which, under successive regimes has sought to impose its authority on Bretons, Basques, Flemings, Alsations and the peoples of the langue d'oc. The Third Republic could extol the wonder of the French nation and its schoolteachers recount its long march of self-discovery; they could not hide its essentially artifical nature. Its northern and southern borders were not secured until the seventeenth century, Nice and Savoy were added as recently as 1860 and Alsace and Lorraine were Part of Germany before 1870 and 1918 and between 1940 and 1945. It was not until after 1870 that the Third Republic was able to turn 'peasants into Frenchmen' (Weber, 1970) and local dialects and customs persisted even later. In the post-war years, France, like several other European states has witnessed a reaction to these assimilationist efforts in the form of a reassertion of the submerged nationalities or micronationalités as they are described in French to distinguish them from the overarching nationality of France and the other 'nation-states' of Europe. Three sets of concerns have contributed to this repoliticisation of the periphery: a cultural revival and reassertion of traditional values in the face of the assimilationist pressures of modern capitalist development and of the state; a sense of economic grievance finding expression in territorial as opposed to purely class or sectoral terms and focussed initially on the state itself; and a movement for political autonomy or 'home rule'. We can trace, in France as elsewhere (Smith, 1981), a progression on the part of regionalist movements from traditionalist, often backward-looking cultural concerns to a concern with contemporary economic problems. This itself may, at least initially, take a defensive form, seeking protection for activities threatened by industrial modernisation; or it may give birth to modernising forces seeking economic transformation. We can trace a further progression, from a strategy of territorial lobbying focussed on the central state to one of autonomism as movements have gained in maturity and confidence and the expectations raised by the central state have been dashed. The relationship amongst the cultural, economic and political aspects of regionalist movements is complex and often bewildering. In purely rationalist terms, one might expect declining regions, needing state help to rescue their economies and preserve living standards, to favour centralised government combined with redistributive regional policies. In particular, one would expect this demand to be taken up on the left, with its commitment to progressive redistribution and equity; and to a great extent this does happen. However, the regionalist dynamic set in train by redistributive economic demands can and does fuel autonomist political movements, paradoxical though this may appear (Lijphart, 1977). A successful regionalist movement indeed needs to integrate these various elements into a mobilising ideology based upon the claims of a territorial or ethnic group. It is their success in achieving this together with the reaction of the central state which explain the fortunes of regionalist movements. In mainland France, as we shall see, micronationalist movements have failed to integrate the elements and relate them to identifiable territories. Consequently, autonomist movements have been weak and fragmented, unable to develop economic and political strategies which could convincingly be presented as alternatives to the centralisation of the state or modern capitalism, or to sever the partisan and clientilistic links tying the peripheral regiona to Parisian politics.

THE GROWTH OF MICRO-NATIONALISM

Regional sentiment never died away in France but in the nineteenth and early twentieth centuries tended to be associated with the clerical and reactionary right, from the

Provençal Poet Mistral, who had refused to support the struggles of the nineteenth century vinegrowers to the Vichy regime, which encouraged provincialism as part of its philosophy of a return to the land and traditional values. The 1950s and 1960s, however, saw a reawakening of the periphery, a new emphasis on economic as against cultural aims and a shift of the regionalist issue from the right to the left of the political spectrum. Social and economic change was transforming provincial France. Old divisions between clericals and anti-clericals were becoming less acute. Modernisation and increased national and international competition were posing a threat to traditional industrial and agricultural sectors, especially in regions remote from the main markets of Europe. The entry of outside capital into the regions was both a boon, in providing jobs, and a threat to local economic control. Following the early example of the Breton CELIB, modernising forces vives, new, dynamic social groups including industrialists, trade unionists, modern farmers, professionals, public servants and academics, began to organise themselves in comités d'expansion économique. Challenging the traditional notables and the centralising state, these organised themselves on a regional basis. The region and, below it, the pays, were seen as providing units both better suited to the needs of modern planning and administration and more attuned to popular identification than the départements. Such thinking, and the associated view that France could be built from below, on the basis of 'natural' units was of course anathema to the jacobin view of the indivisible republic to be divided into manageable units for administrative purposes only. It would be wrong to generalise about these movements; the extent and significance of activities of forces vives varied considerably. In some regions, movements were forged pushing for modernisation and development; in others, there was little more than a defence of archaic economic structures. While the Fourth Republic governments had broadly welcomed the comités d'expansion, the reaction of the Gaullist state was to try and control them, to subordinate them to national policy (Hayward, 1969a) through the Commissions de développement économique regional (CODER), established in 1964 (see Chapter Two). The attempt to co-opt the forces vives to the service of the state, so outflanking the notables, still tied to fourth republic politics, largely failed and many of them resigned from the CODER when they became aware of their impotence. Following the failure of de Gaulle's further attack on the notables in the abortive regionalision measure of 1969, the Pompidou government temporised. The indirectly-elected regional councils established in 1972 effectively excluded the <u>forces vives</u>, co-opting, instead, the <u>notables</u>. Instead of being the instrument of modernisation, regions were now to be a buffer to mitigate its consequences (Gremion, 1981). Regional boundaries followed closely those of the former <u>circonscriptions d'action régionale</u>, deliberately drawn so as not to correspond to cultural identify or historic Provinces - Brittany, notably, was cut into two. Nor, for that matter, did they correspond well with the technical needs of economic and physical planning. So while some regional councils showed considerable dynamism in their economic and cultural development strategies, they did not become the focus of ethnic or micronationalist activity.

Regional policy initiatives, too, were developed in a highly centralised way which merely served to heighten regional consciousness without increasing the legitimacy of the central state. In 1963, the <u>Délégation à l'Aménagement du Territoire et à l'Action Régionale</u> (DATAR) was established to give a regional dimension to national planning and encourage the dispersal of industry to the regions. Regional development became a favoured phrase and was used to justify a variety of projects. But, as Tarrow (1978, p. 99) comments, 'when regionalism as ideology is used by political elites to justify policies that have little to do with regional equality, regional political defense - in part touched off by the ideology - may result.' Gremion and Worms (1975) similarly comment that regional policy may have 'contributed to an increase in regional consciousness which is at the outset awareness of frustration.' The case analysed by Tarrow (1978), was the notorious Fos Project undertaken through DATAR for a steel complex at Fos new Marseille. Although justified by reference to the needs of regional development, the project was ill integrated into the local elites, it was taken over by the centre and turned to the needs of national and European development (Paillard, 1983). The result was a coalition of regional defence, comprising local politicians, territorial administrators, environmentalists and trade unionists in opposition to the scheme. Other large-scale but isolated industrial development initiatives have been dismissed as 'cathedrals in the desert'. Similarly, the massive tourist developments on the Mediterranean, undertaken through the Parisian ministries and mixed public-private companies, while they have generated employment, have not been viewed as an unmixed blessing. Local politicians have complained that the profits have gone to outside property companies and there is some animosity to the annual influx of strangers (Ardagh, 1982). There is resentment too, about the seasonal nature of the work and the subservient attitudes associated with the tourist trade. These attitudes have fed a feeling that the region is being exploited by the state for its own ends and to calls for more decentralised and planning, sensitive to local needs. Langumier (1983) sees the regions as having been subordinated to the three successive imperatives of the fifth republic state, gaullist modernisation of French society, pompidolian industrialisation and giscardian redevelopment in the interest of European and international competition. Policies have not, according to this analysis, been geared to the needs of the regions themselves.

At the same time, there has been a cultural revival in the French periphery. While it is estimated that there are only about 600,000 Breton speakers left, there has been a revival of interest in the language among the young. Berger (1977) cites figures showing an increase from under 200 to nearly 800 in the number of students offering Breton in the baccalaureate examination at the Academy of Rennes. Elsewhere, the contempt for <u>patois</u> among the upwardly mobile is disappearing and the study of regional languages has been increasing. The state has been pressed into allowing their use to a limited extent in education and broadcasting in a series of measures starting with the Loi Deixonne of 1951. Lirerary output varies with a thriving Breton literature but less of a flowering elsewhere. Folkmusic and dance have also featured in the cultural revival and in the 1970s there even developed a Breton rock with considerable mass appeal among the young.

The modernisation of the periphery, the emergence of the <u>forces vives</u>, the cultural revival, together with the clumsy attempts at centrally-directed regional policies, have provided the conditions for social movements in the regions of France. Their political development was helped by three crucial sets of events, the Algerian war and independence, the 'events' of May 1968 and the party political developments, particularly on the left. These were given a political opening in the party realignments of the 1960s and 1970s. Algerian independence dealt a severe blow to concept of the one and indivisible republic; having set its boundaries in North Africa, the political class had to accept decolonisation as dismemberment, making it more difficult to sustain the myth (Beer, 1977). May 1968 gave voice to <u>autogestionnaire</u> (self-management) ideas, questioning the centralisation both of the regime and of much left-wing thought. Environmentalists and decentralists could make common cause with the 'small is beautiful' philosophy against the gigantism and inhumanity of both modern capitalism and the state. The defence of territories and traditional cultures could be seen as part of the same cause. All this amounts to, so far, however, is a collection of 'good causes' without an integrating ideology or programme. It was to provide this that regionalist intellectuals developed the 'internal colonialist' thesis, rooted in Gramscian analysis and brought to France by the Occitan nationalist Robert Lafont. According to Lafont (1967), internal colonialism is an instrument of capitalist domination, to prevent the emergence of either class or regional consciousness. From the nineteen century, the industry of Languedoc and other regions was taken over by external interests, with the collaboration of the native bourgeoisie and reduced to a dependent mode. Indigenous agriculture was displaced and local distribution circuits replaced by external control Extractive industry is encouraged, not for the benefit of local industry but to provide materials for elsewhere. Nationalisation is no solution for the state itself has, as depicted by James Burnham, been taken over by the managerial classes aligned with the needs of capitalism. So 'technocracy and neo-capitalism have seized the state and together, without any intervention by the citiens, choose the great options on which depend the whole future of the country' (Lafont, 1967, p. 63). The regional policy measures taken by the state are merely the equivalent of the neo-colonialism which has replaced overt colonialism in the Third World. At the same time, regional cultures are reduced to mere folklore, spectacles for the tourist industry.

Liberation, according to Lafont, will come only when cultural and economic concerns are linked. Yet, in the past sympathetic economists had supported regional autonomy and indigenous development but regarded regional languages as reactionary aberrations. Some cultural regionalists, on the other hand, had not wanted to see economic issues getting in the way of their primary concerns. The nationalist sees cultural and economic subordination as two facets of the same phenomenon.

There are many intellectual problems with this analysis. For example, is the Parisian working class to be considered among the exploiters or the exploited? If the latter - which is presumably what Lafont and his colleagues would argue - then the exploitation ceases to be primarily, if at all, a territorial one. Further, the internal colonial model as developed elsewhere (Hechter, 1976) is predicted on the notion of a cultural division of labour. Lafont emphasises only a territorial division of economic sectors, which is not necessarily to be explained in terms of colonial exploitation. The importance of the theory, however, lies less in its academic rigour than in its ideological potential. It is able to link cultural with economic concerns, relating both to definable territories, and to link socialism with decentralisation. It further gains plausibility by the fact that capitalism does develop in an uneven manner and that industrial concentration, in France as elsewhere, has taken control of industry out of the hands of provincial owners.

The internal colonialist thesis was taken up by Michel Rocard's <u>Parti Socialist Unifié</u> (PSU) at their <u>assises</u> of Grenoble in 1967 (Rocard, 1967) and has enjoyed considerable currency amongst the regionalist and <u>autogestionnaire</u> groups and parties which flourished between the demise of the SFIO and the rise of the <u>Parti Socialiste</u>. The demands which stemmed from the new analysis were for a political decentralisation of a federal type together with regional and local control over the levers of economic power. The slogan was '<u>vivre, travailler, decider au Pays</u>'. The political programme for the regionalist revolution, however, was never clearly spelt out and there was a profound ambiguity in the demands for radical measures of Home Rule. Lafont himself recognises that the really big issues will have to be decided nationally and even betrays his French prejudices to the extent of including among these the national education system! Internal colonialism may be an attractive idea to the intellectuals but popular support for regionalism rested on shifting and fragile foundations. We can illustrate this by looking at two regions, Brittany and Languedoc.

BRITTANY

Brittany appears to have most of the prerequisites for a micro-nationalist movement. Formerly an independent duchy annexed to France in 1532, it has a language and culture of its own and, as a peninsula, a clear territorial identity. Traditionally, it has been a very Catholic region and has voted overwhelmingly with the right, though, with the decline of religious observance, both of these characteristics are changing (Johnson, 1981). Indeed, it was the Church that was largely responsible for keeping the language alive under the Third Republic, seeing in it a defence against the centralising secular state - though left-wing Bretons now claim that the Church stifled the development of the language by encouraging local dialects as a means of keeping the masses in ignorance. By 1950, Breton nationalism, discredited by the wartime collaboration of some of its activists, was at a low point; Breton speaking was beginning a rapid decline with the advent of television (Stephens, 1976). it was in this year, however, that the CELIB (<u>Comité d'études et de liason d'interêts bretons</u>) was founded by a coalition of progressive business, agricultural academic and, in due course, trade union leaders (<u>forces vives</u>), to press for the adoption and financing by the government of a Plan for Brittany. So the Breton issue was brought back onto the agenda in a respectable form, tied to the needs of economic development. CELIB's strength and its weakness was the broad base of its support, from local government, members of Parliament, industrialists, trade unions and farmers. The Communist Party for a time opposed the movement as an example of 'class collaboration', but the Communist-led CGT union finally came in in 1961. Regional solidarity may paradoxically have been helped by the cohesiveness inherited from the traditional Catholic 'counter-society' in Brittany (Gremion, 1981) as well as the growth of secularisation which reduced the traditional divisions within Breton society. The modernisation and secularisation of Breton society further dispelled the fears of the left that any concessions to Breton identity would be to the benefit of superstition and reaction. Solidarity was consolidated around a series of economic struggles, starting with the campaign for the regional plan and continuing with the long fight to keep open the <u>Forges d'Hennebont</u> and the successful <u>bataille du rail</u> to force the nationalised railways to withdraw a proposed tariff change which would have disadvantaged Breton traders (Dulong, 1975). A plan for Brittany was adopted by the state - though the finance necessary for its implementation was not guaranteed - and the governments of the Fourth Republic were induced to embark on a series of regional policy measures (Phlipponneau, 1982). The high point of CELIB's influence was in 1962 when it succeeded in extracting from most of Brittany's candidates in the general election a pledge to put the interests of Brittany above those of party (Hayward, 1969a). The Breton electorate, however, was still inclined to the right and, with the election resulting in a large Gaullist majority, the reverse occurred. Rene Pleven, the President of CELIB, accepted government office. The establishment of the CODER in 1964 largely succeeded in its aim of emasculating the movement, channeling regional energies into a centralist device subordinate to the new regional prefect. These developments led Michel Phlipponneau, a Rennes professor and leading Breton actiist, to conclude that the only hope for the Breton movement was to throw in its lot with the left. In 1965 he joined Gaston Defferre's abortive campaign for the Presidency, inserting into his programme a pledge on regional reform. Unable to persuade the movement as a whole to join him and excluded from the CODER in 1964, Phlipponneau resigned from CELIB in 1967 complaining of a <u>trahision des notables</u>. In 1972, the CGT left.

While the CELIB continued to put its faith in a strategy of non-partisan pressure on the central state, other sections of the Breton movement veered to the left, a shift which was accelerated by the events of May 1968 when the anti-state protests struck a familiar chord in Brittany and by a series of industrial struggles. In 1972, a strike at the Joint Francais for, among other things, parity in wage rates with the Parisian workers of the same firm, generated a considerable amount of regional solidarity (Berger, 1977); but it was a solidarity of the left rather than the cross-class mobilisation represented by the CELIB. Other movements around the same time, included a protest by small milk producers against their co-operative which they saw as treating them as would any capitalist dealer, and the demonstrations against plans for military bases.

The Breton cultural movement, too, saw a revival in the late 1960s and early 1970s. In 1970 Breton was admitted as a subject for the <u>baccalauréat</u> (Stephens, 1976) and the number taking it increased steadily as did the numbers of correspondence courses in the language. By 1976, there were some 20 periodicals in Breton and, despite the small market, novels, plays and other works were being produced (Stephens, 1976). The Church was also reverting to its traditional policy, encouraging the use of Breton in the liturgy. None of this, however, amounted to a national revival. Numbers studying Breton remained tiny in proportion to the total population. Daily use of the language was continuing to decline and, as one observer acidly commented, 'parents who speak Breton to their children today do so deliberately and are therefore militants' (quoted in Stephens, 1976, p. 400). While this may have been a weakness from the point of view of Breton micronationalism, it did ensure that the emphasis of the movement remained on economic affairs and thus in the political mainstream. A heavier emphasis on language could have been divisive, as in Wales. More significant at the

popular level was the revival of Breton music and its
translation into modern idiom by singers such as Alan
Stivell whose electric folk-rock is used to convey his pan-
Celtic and Breton message. The folk revival in Brittany,
as elsewhere, had particular appeal to the young.

This was the background to the development of a more
politicised Breton nationalism, aiming at decentralisation
and home rule. As early as 1956, the <u>Mouvement pour
l'organisation de la Bretagne</u> (MOB) had been formed, calling
in a manifesto signed by several thousand for
decentralisation of the French state, a special statute for
Brittany with an elected assembly, as well as immediate
economic measures. Some of its members were elected to
municipal councils in the 1950s (Stephens, 1976). In 1963,
it split following the defeat of a conference resolution
calling for Algerian independence, to give birth to the
<u>Union democratique bretonne</u>, a political party, socialist
and internationalist. Campaigning on the internal
colonialist slogan 'The Breton worker is doubly exploited;
he is exploited as a Breton and as a worker', and calling
for a European federation of regions, the UDB formed
tactical alliance with other left wing parties and several
of its members were elected to municipal councils on union-
of-the-left lists (Guillorel, 1981). The UDB, however, was
not to be entirely successful in marrying socialist and
regionalist ideologies and even less so in building a mass
base. In 1968 it expelled those members sympathetic to the
violent FLB and in 1970, after moving to a strict Leninist
line, expelled its <u>gauchiste</u> libertarians and aligned itself
with the Communist Party. Nevertheless, by the 1970s there
was a firm association of Breton regionalism with the left,
given added force by the advance of the Socialist Party in
Brittany, using as one of its weapons a firm commitment to
decentralisation. Morvan Lebesque (1970) recalls how his
left-wing Parisian friends used to tell him that Breton
nationalism was reactionary and wondered how he could
interest himself in both Vietnam and Brittany. In the
aftermath of 1968, however, with decentralised socialism in
fashion, this no longer seemed so odd.

The MOB, left with the right-wing rump of the nationalist
movement, transformed itself in 1972 into <u>Srollad ar Vro</u>
(SAV), with the slogan, 'neither red nor white, only Breton'
but this, after oscillating between extreme right and
extreme left (Phlipponneau, 1982) has virtually disappeared.
In the late 1960s and early 1970s a series of terrorist
groups sprung up, attacking the symbols of 'French
occupation.' The most prominent was the <u>Front de</u>

libération de Bretagne whose campaign began in 1966. From 1969 to 1972 there was a series of attacks by the Armée révolutionaire breton. While the numbers of people involved in these activities was small, there were indications that the Breton water was not totally hostile to the guerrilla fish. A wave of arrests in 1969 netted some very respectable citizens who were never brought to trial but released under an amnesty by the new Pompidou government. In 1972, after another series of arrests, leaders of the legitimate Breton groups took the attitude of deploring the methods of the accused but understanding their aims and the frustration which drove them to desperate acts. Such a tolerant attitude was helped by the fact that the targets of the attacks were property rather than persons, though this did not save the FLB from being banned in 1974.

By the 1970s, the politics of most Breton activists were within the democratic socialist fold, enabling them to contribute to the realignment of the left. Political clubs such as the Bonnets rouge and Bretagne et Democratie and conferences on the lines of the Grenoble assises of 1966, at Saint Brieuc and elsewhere (Phlipponneau, 1982) paved the way for the adoption of regionalism by the Parti Socialiste, as we shall see. From the agreement on the Common Programme of 1972, the main effort of the Breton regionalist movement was channeled into the parties of the left, which had taken on board the substance of their demands for regional government, local economic powers and protection of Breton culture. This accommodation to Breton regional interests coincided with a steady advance for the left in Brittany, from 26% in the legislative elections of 1967, to 33% in 1973 and 43% in 1978 (Frears, 1978; Rogers, 1984). At the legislative elections of 1981, the left won 15 of the twenty five Breton seats. At local level, the Socialists took control of the conseil général of Cotes du Nord in 1976 and in 1977 gained the mairies of Brest, Rennes and St. Malo.

LANGUEDOC

The micronationalism of Occitania or Languedoc is even more diffuse and fragmented than that of Brittany. There is endless controversy, indeed, over what the boundaries of Occitania are or should be. While the two names, Occitania and Languedoc have the same derivation, it is usual now to distinguish between the widely defined Occitania and the narrowly defined Languedoc. On a wide definition, Occitania covers a large part of France, taking in Provence, Limousin, Auvergne, Gascogne, Guyenne and Dauphine as well as the Mediterranean lands of Languedoc (Stephens, 1976). The regional reforms of 1963 and 1972 furthered the confusion by creating the region of Languedoc-Roussillon, combining parts of historic Languedoc with French Catalonia and areas of Provence but allocating Toulouse to the region of Midi-Pyrenees. On both the wider and the narrower definition, the region is distinctive culturally. The langue d'oc, a romance tongue distinct from French, thrived in the Middle Ages, the golden age of the troubadors, when Occitan literature flourished. Its decline started as early as the fourteenth century when, with the advance of French political power, the Occitan elites went over to the French language (Weber, 1977) and the langue d'oc came to be regarded as a series of peasant patois. Unlike Brittany, the South tended to favour the revolution and the nineteenth century saw strong support for secularisation and republicanism - and thus francophone hegemony - among the local political leadership. A cultural revival was started in the late nineteenth century by the Poet Mistral and his colleagues in the Félibrige but, when not openly contemptuous of political activity, these tended to side with the traditional right-wing and monarchist forces.

For our analysis, we shall concentrate on the narrowely-defined area of Languedoc, which is where the main micronationalist activity has occurred. The Third Republic saw the emergence of a political structure based on the ability of notables of the Radical Party to manage the relationships of the dominant vine-growing community with the state. The result was increasing state support for the monoculture of vine-growing and a clientilistic political system which, in the twentieth century was taken over from the Radicals by the socialist SFIO. Economically conservative but politically 'left-wing', the peasant vine-growers were highly dependent on the support and protection of the state which continued into the post-war period, despite recurrent crises of over-production and disease.

A political crisis, however, developed from the 1960s when the state, under pressure from EEC rules, began to cut back on support for viticulture while at the same time sponsoring measures for agricultural and industrial diversification. We have already mentioned the Fos affair. Another example was the Libaron development (Lafont, 1967). This arose from the inability of fruit producers in Languedoc to sell their produce on the saturated European markets. The solution was to go for conserves. So, at the initiative of the state, the multinational giant Libby's were brought in and, despite protests from local co-operatives and farming unions, were soon put in a position to dominate production as well as marketing, but without guaranteeing to tie itself for the future to regional sources. The canal development of the 1950s and the tourist developments from the 1960s also aroused suspicion amongst peasants and the notables, who saw their power threatened. Viticultural discontents came to a head in the 1970s; a series of riots over Italian wine in 1975 left two dead.

The cultural renewal emerged from the Resistance after the war in the form of the Institut d'Etudes Occitanes. Campaigning essentially for the revival of Occitan culture within the French nation, it was initally divided on its political aims (Touraine, 1981). Certainly, it had a more modernist image than the Félibrige, which it saw as provincialist, conservative, clerical and defeatist; the Félibrige in turn accused the IEO of being Marxist and revolutionary (Stephens, 1976). At the same time, there was something of a revival of Occitan poetry and song, though little by way of prose or drama.

In 1959 there was the beginnings of a politicisation of the movement. A small group broke away to form the right-wing Parti Nationaliste Occitan, while the mainstream of the IEO moved left, adopting a policy in favour of political decentralisation. In 1961, a strike by miners at Decazeville against a closure threat generated considerable regional solidarity and gained the support of the IEO. A new movement, the Comité Occitan d'Etudes et d'Action (COEA) emerged, bringing together the cultural and economic themes in an 'internal colonialist' interpretation of Occitania's plight (the term 'internal colonialism' had been first used by Robert Lafont and others in a manifesto of 1961). It was opposed to the state's strategy of aménagement du territoire which it saw as a means of keeping the south in a state of permanent dependence and under-development and favoured federal government. It gained the support of a number of left-wingers who despaired of achieving power at the national level and in 1964, it was one of the founders of Francois Mitterrand's Convention des Institutions Republicaines (CIR). May 1968 sparked off a new cultural movement linked to the more anarchic sections of the left and uniting peasants, intellectuals, gauchistes, trotskyists, ecologists, pacifists and anti-imperialists (Touraine et al. 1981). In 1971, this led to the dissolution of COEA in favour of Lutte Occitane, a more militant group which had effectively taken over COEA and which led to series of protests against the Larzac military base in 1971-2. In 1974, a new campaign of agitation was being planned when the death of President Pompidou provided a political opportunity. Lafont was put forward as presidential candidate of the 'minorities and regions' but his nomination was rejected by the Constitutional Council. With the opportunity to enter the partisan arena thus closed off, the movement aligned itself with the existing parties of the left. The slogan of Lafont's abortive campaign Volem Viure al Pais (VVAP), became the name of a new organisation favouring regional decentralisation within the French state and giving general support to the parties of the left. By the late 1970s, it had persuaded both the Parti Socialiste and the Communists to support its main demands, though their local notables are not always so favourable (Carelli, 1980). This was made easier by changes within these parties themselves, especially the renewal of the old SFIO as the Socialist Party, which we examine in the next chapter. A new generation of younger activists, professionals and public sector workers especially, were coming to challenge the power of the old notables and gaining election to local councils in the 1970s. While the old notable might declaim against the state in the name of the locality, his relationship with it was in fact symbiotic, his power deriving from his role as the privileged channel of access. Now this role was being undermined by the inability or unwillingness of the state to deliver the traditional goods and by the emergence of the new breed of activist more committed to genuine decentralisation and a more politicised mode of decision-making. The Communist Party, traditionally very centralist, changed tack during the 1970s. While rejecting the 'internal colonialist' argument as unmarxist (Giard and Scheibling, 1981), it has supported Occitan cultural ventures and taking a strong line in favour of vine-growers and against EEC enlargement. For the European elections of 1979, it recruited to its list the unlikely figure of Emmanuel Maffre-Bauge, landed leader of the local viticulturalists.

Occitan regionalist agitation reached its peak in the mid 1970s, under the impetus of the viticultural protests, the Larzac campaign, and the cultural revival. The strategy of cooperation with the French left, however, was a two-edged sword. While providing a realistic prospect of achieving results, it did tend to demobilise the movement as an independent force. The break-up of the union of the left in 1977 and the subsequent defeat of the left at the 1978 legislative elections came as a severe blow producing two contradictory tendencies, to unity and to disintegration. A series of attempts was made to unify the former activists of Lutte Occitane and VVAP, culminating in a meeting in 1979 to approve a manifesto and constitution for a new body to be called Movement Socialista Occitan - Volem Viure al Pais. The meeting failed to bridge the gap and Lafont and other left-wingers left VVAP, aligning themselves again with the French left in the period up to the 1981 Presidential elections. VVAP, for its part, moved increasingly in the direction of fundamentalist nationalism, with attacks on the French left as well as the right. The old Lutte Occitane activists continued to publish their journal for a while but then dropped out of active involvement. VVAP's strategy of complete independence yielded meagre results. At the cantonal elections in Spring 1985, three of its four candidates received derisory votes of less than 3.5%. The fourth, the movement's president and mayor of a small commune, came second with 18.2% in the first round and in the second round, with the support of the eliminated left wing candidates, got 46.5% against the candidate of the UDF. Ironically, the result appeared to vindicate the strategy of cooperation with the left to which VVAP was in general opposed, but the performance of the Socialists in government had by now created too much disillusion for this to be resumed.

The point emerges more strongly still when one examines the support of Occitan regionalism among other groups and political movements. While, as some writers have pointed out (Barelli et al. 1980), the Communist Party and its associated trade union, the CGT, supported the campaigns of economic defence and were prepared to take up Occitan themes, the mainspring of their strategy was for regional protectionism within a protected French market. Their vision of regional government was a limited one, firmly located within the French state and hostile to the European Community and a fortiori, its enlargement. The CFDT trade union was more positive in its commitment, in line with its autogestionnaire and decentralist ideology, and included a number of Occitanists in its ranks. In the Socialist Party, there was also support for regionalism as, in the transition from the old SFIO to the new Parti Socialiste, much of the traditional jacobinism had been shed. The new regionalists, however, co-existed with the old centralists and the notables, rooted in the institution of the départements and were themselves divided between sympathisers with Occitan nationalism and technocratic modernisers who saw the region as a framework for planning and development. The Socialists' identification with the viticultural movement was tempered by their strong pro-EEC position, in contrast to the Communists.

It is not surprising, then, that groups like
Lutte Occitane and Volem Viure al Pais approached the
question of collaboration with the French left with some
circumspection. Lutte Occitane summed up their position in
the aftermath of the 1978 elections (Pais Occitan, 33 Prima,
1978). They had hoped for a left victory because they
thought that it might amount to more than a mere change in
the personnel of the state. Regional autonomy as promised
by the left might be the first step to real autonomy for
Occitania. The social programme of the left was in the
interest of Occitan workers; and, if the left would not
solve their problems, the right would merely make them
worse. On the other hand, they were aware that the left's
recognition of Occitan culture did not amount to a
recognition of their own socialist and autonomist project;
and that the productivist model of the left was far from
their ecological, autogestionnaire, anti-nuclear views.
From the mid 1970s, as we have noted, Lafont himself worked
increasingly with the left, helping to frame their
regionalist programmes, in an effort to synthesise Occitan
and socialist themes, while VVAP moved to a more independent
position. By 1980, Lutte Occitane could note that hitherto
'cette strategie d'integration cironspecte a la gauche etait
egalement celle de VVAP avec les confusions propres a ce
type de rassemblement qui regroupe a la fois des
'nationalistes' occitan et des militants de partis francais
- certains reussissant le tour de force d'integrer cette
contradition en une personne' (Alcouffe, 1980). The person
referred to may have been the Occitan poet Yves Roquette who
confuses our analysis by being both a fundamentalist Occitan
nationalist and a member of the French socialist party
(Roquette, 1982)!

The tussle between fundamentalist nationalists and those
looking for a transformation of the French state breaking
down barriers between people in the context of a more united
Europe is a constant theme. The Parti Nationaliste Occitan
stands at one extreme, accused by its opponents of racism
and chauvinism. VVAP was for long torn between the two
position and, with the departure of Lafont and his
associates, increasingly adopted the fundamentalist posture.
By 1985, it was calling for the elimination of tourism, a
strategy of autonomous development and priority in jobs and
training for people born in the region. Such a view was
not to the taste of people like Lafont who wrote as early as
1972 that 'L'independence n'est pas l'autonomie.
L'independentisme est un romantisme qui a une certaine
importance par moments mais l'autonomie prospective des
regions doit articuler diverses autonomies' (Lafont, 1972).

In other words, Occitanism must not become an exclusive focus of loyalty as French nationality had but must encompass a more pluralistic view of the world.

At the same time, emphasising the primacy of Occitan issues entailed a downgrading of socio-economic divisions and a denial of any significance difference between the French left and right. So, while VVAP still describes itself as socialist, its opponents claim that its attacks on the left parties place it on the right. Its rather indiscriminate sympathy for any 'national liberation' movement, including ETA, the IRA and Fleming extremists also marks it off from the old <u>Lutte Occitane</u> activists who regarded the ETA campaign, for example, as unacceptable now that Spain is a democracy.

Drawing these elements together, then, we can see a contrast between a tendency to a fundamentalist, inward-looking, conservative and defensive nationalism on the one hand and a left-wing, outward looking, cosmopolitan regionalism which places Occitania in the context of a Europe of the regions in which nation states are reduced in importance. The same tensions are evident in the cultural field, with, on the one hand, the <u>Félibrige</u> tradition, romantic and <u>passéiste</u> and, implicitly at least, conservative and, on the other, the movement for a modern, living Occitan culture. Social and demographic changes in the cities, however, have made them infertile ground for the latter and, in contrast to Brittany, there has been no Occitan youth culture or rock-music. Without a base in the expanding cities occitanism thus retains a rural image, one of a pre-industrial and to some degree anti-industrial movement. Yet, despite the relative weakness of Occitan culture, writers have frequently been to the fore in the regionalist movements, discouraging the development of serious economic and political thought.

REGIONALISM IN THE 1980s

We have seen how micro-nationalism in Brittany and Occitania developed, bringing together cultural, economic and political claims to produce social movements of differing strengths and developing to different points. Modernisation and secularisation were by this time beginning to break down the old view of the provinces as hotbeds of reaction and clericalism so allowing a rapprochement between regionalism and the left. At the same time, crises of economic and political representation were undermining the old <u>notable</u> system while the realignment of the national party system was allowing new issues onto the political agenda. Though the micronationalist movements were past their peak by the time of the Socialist victory in the 1981 elections, the Party had, in the course of its evolution, taken on board some of their demands. The party's growth and the break with the old SFIO style of notable politics provided an opening to various sections of the non-traditional left including the regionalist movements which, as we have seen, were themselves moving in a leftward direction as well as the political clubs, usually more favourable to regionalism (Phlipponneau, 1977) and, eventually, Rocard and most of the PSU. The espousal by the Parti Socialiste of most of the regionalists' demands undercut support for the one-issue groupings but only, as we shall see, at the cost of introducing considerable ambiguity into the Socialist programme itself.

In the longer term, the prospects for micronationalist movements in mainland France remain constrained by the factors which have made for their weakness in the past. Except perhaps in the case of Brittany, there is a weak sense of territorial identity and a lack of a tradition of independent existence. We have seen how the state has deliberately maintained this lack of territorial cohesion; and the traditional pattern of territorial representation has proved remarkably resilient. There is the difficulty of integrating cultural and economic demands and the diverse and changing nature of the latter. There is the difficulty of integrating the concerns of the diverse classes and occupational groups to which the movements seek to appeal - from dynamic forces vives to conservative peasants, from industrial trade unions to radical ecologists. The movement of regionalism to the left has to some degree resolved certain of the ideological difficulties by identifying it with the forces of social and economic change rather than as a class and ideological catch-all. On the other hand, this places it in the dilemma of other movements seeking to tap both nationalist and socialist support (Keating and Bleiman, 1979). If they seek to work through the national party system, they will find their own parties always have higher priorities. If they form their own parties, they face isolation from the political mainstream and a loss of influence at the centre and, unless the regional electorate are prepared to put that issue at the top of their agenda, electoral irrelevance, especially in states with non-proportional electoral systems like those of Britain and (up to 1986) France.

Then there is the crucial relationship between territorial

lobbying and support for political decentralisation. As we have seen, the dynamic of regional movements in France as elsewhere has led them beyond lobbying the central state towards home rule strategies; but for this to be convincing there needs to be some integration of the two strands at the ideological level and at the level of political mobilisation. The unsound 'internal colonialism' argument has proved of relatively little value beyond the ranks of political militants. The socialist policy of developing regionally-based economic sectors and a greater degree of planning of the regional implications of economic developments, while recognising that major decisions are going to be taken elsewhere, may have more appeal. There is still a lively debate about local and regional economic initiatives which is likely to receive fresh impetus when, at last, the regional councils are directly elected. Economic transformation and changing electoral fortunes, with an increase in party competition in regions previously dominated by one grouping (the right in Britanny, the left in Languedoc) will inhibit a re-emergence of the old <u>notable</u> pattern of territorial representation. The new regional leaders on both right and left are more committed to genuine decentralisation and regional autonomy. This, though will be a secular, modernised regionalism, shorn of much of the traditionalism and cultural impetus of the past. One of the key questions to be addressed to the Socialists' actions in government is whether they have succeeded in addressing this issue.

4 Decentralisation and the Parti Socialiste

CENTRALISATION AND THE FRENCH LEFT

It is common to associate socialism with a political outlook which deprecates territorial divisions and supports the centralisation of political power and there are, indeed, some sound theoretical and practical reasons for this. Marxist theories of socialism give primacy to the class struggle, and the capture of power by the working class. From this point of view the politics of territory is an illusion; at worst a diversion from the class struggle, at best a disguised form of it. Class solidarity within and beyond the nation-state is the only true expression of the workers' interests. On the form of state itself and the uses of political activity centred on it, Marxist theory has been profoundly confused and ambiguous. Marx himself at various times expressed the views that the state was merely the executive committee for managing the affairs of the bourgeoisie; and that, at crucial junctures the balance of class forces could be such that political action through the state could tilt it in favour of the proletariat (Miliband, 1977). In later life, he conceded that, in certain countries, socialism could even arrive through peaceful, parliamentary means (Miliband, 1977: Feuer, 1959). Given this ambivalence on the institution of the state itself, it has in practice been possible for Marxists to take up a variety of positions on the legitimacy of local political activity and on the decentralisation of the state, though retaining the emphasis on class solidarity as their fundamental principle.

If for Marxists, politics and the state are often a secondary item to the industrial class struggle, in the non-marxist, 'social - democratic' tradition, they are central. It is by gaining control of the state that socialists can counterbalance the power of private interests, plan the economy in the interests of all and redistribute its fruits in the interests of the deserving. The commitment to equality of treatment for all implies a strong and centralised state, with little room for territorial variation or dispersed power. In socialist parties with roots in the organised labour movement, these centralist leanings are reinforced by traditions of solidarity and class cohesion forged in industrial action. Trade unions, seeking uniformity of wages and conditions across the national territory - to prevent undercutting by employers - are themselves a major influence for centralisation, especially in the public sector.

In France, we can add to these general influences the historical experience of republicanism and the jacobin tradition. The revolution, refounding the state on the principle of popular sovereignty, gave it a democratic legitimacy - at least outside the ranks of Marxists, who tended to assert that the revolution had transferred control of the state from the hands of the monarchy into those of the bourgeoisie. For much of the nineteenth century and especially during the Third Republic, the centralised state was seen as an instrument of progress against the superstition, ignorance and reaction to be found in the localities. Nineteenth century republicanism left many marks on the French socialist movement which arose out of it in the early part of this century, anti-clericalism and jacobinism being two of the most notable. The anti-clerical struggles and the Ferry laws (1879-82) establishing universal secular education are seen as triumphs for revolutionary jacobinism even by socialists who now favour decentralisation.

Not all of the traditions of the French left, however, are centralist. Anarcho-syndicalist and anarchist movements have continued as a minority stream, presenting a vision of decentralised and stateless socialism far removed from the centralising bureaucracy often associated with social democracy. Anarcho-syndicalist ideas had considerable influence in the French union movement around the time of the First World War (Cobban, 1965) but they waned in the inter-war period. After the Second World War, with the major union, the CGT (Confédération Générale de Travail) under the effective control of the Communist Party, syndicalist ideas were given little scope. It was the 'events' of 1968 which brought their rediscovery. The libertarian socialism of the student revolt and the fashion for autogestion (self-management) were taken up by sections of the union movement, notably in the CFDT (Confédération Française Démocratique de Travail) as well as the minority parties of the left such as the Parti Socialiste Unifié.

We have noted the ambivalent attitudes of the left to the state. On the one hand, it is a potential instrument for progress; on the other, it is regarded by some marxists as necessarily operating in the interests of the bourgeoisie. In France, support is given to the latter view by the tradition of centralised state power on the political right and the close association of the post-war state with big business. Indeed, a distinction is often drawn between the nexus of state and big capitalism on the one hand and popular forces on the other. The latter category is easily stretched to include not only the working class but peasants and small business and peripheral interests generally. So attacking the centralised state, far from weakening an instrument of the class struggle, can strengthen popular forces; while the extension of state power in social and economic matters, so often undertaken by the French right, is by no means necessarily to be identified with socialist advance.

The extension of the category of progressive forces to include peasants and small businessmen is an implicit recognition of the numerical weakness of the industrial working class in France. Though French socialists retain a remarkable loyalty to Marxist rhetoric about the class struggle and the interests of the proletariat, in practice the left has had to build a coalition from such progressive forces as were to hand. In recent years, these have included regionalist and decentralist movements as well as ecologists, autogestionnaires and others with little commitment to the jacobin state.

THE EMERGENCE OF THE NEW FRENCH LEFT

At the end of the second World War, the two main parties of the French left - the SFIO (Section Française de l'Internationale Ouvrière) and the Communist Party - remained firmly attached to the jacobin myth. The SFIO was not only rooted in its republican past. In the present, it feared the advance of the Communists and the use they might make of their strength in local government and tacitly agreed with the right to allow those sections in the constitution of the Fourth Republic vesting the executive power in the départements in the presidents of the conseils généraux to remain a dead letter (Meny, 1974). All traces of the regional organisation left by Vichy were to be swept away, though this proved difficult in practice, and the only local institutions which were to remain were the traditional republican ones of the département and the commune. There was self-interests as well as principle involved here. In the post-war period, the SFIO, losing both popular support and active members, was forced back on its strongholds like the Nord and Marseille (Shaw, 1983). It was dominated by 'middle aged civil servants and middle class professionals' (Johnson, 1981, p. 144) and, above all, had become a party of notables; in 1965, over half its members were local councillors (Frears, 1977, p. 107). Not surprisingly, it maintained a keen attachment to the existing pattern of territorial government. The advent of the Fifth Republic, which most of the party had initially supported, placed the SFIO in permanent opposition in national politics and its vacillating attitude to the Gaullist regime lost it many members.

As for the Communists, they were no less zealous in defence of the local institutions of what was still, in their eyes, a bourgeois state. Self-interest had its part here, too. Communist control of local government, particularly in the 'red belt' around Paris played a vital role in sustaining the counter-culture in which the party could survive. Modernisation of local government, by upsetting established relationships, could make a breach in the political 'ghetto' to which it had retreated with its working class supporters (Johnson, 1981). In any case, the doctrine of democratic centralism left little space for dispersed power, pluralism or dissent and the principles which guided the party in its internal affairs were assumed to be those most appropriate for a revolutionary government, should one ever come to power.

It was demise of the SFIO, the emergence out of it of the new Parti-Socialiste and the forging of the Common Programme with the Communists which allowed the French left to shake off its centralist legacy. This complex story has been told elsewhere, (Johnson, 1981; Frears, 1977; Bell and Shaw, 1983). Here we must limit ourselves to those aspects relevant to our theme. By the early 1960s, it was apparent that the SFIO was in terminal decline. The key question was whether socialists should seek an opening to the centre or should strike a left-wing path and attempt to come to power in alliance with the Communists. Gaston Defferre, the mayor of Marseille, was the leading exponent of the centrist strategy which he attempted to put into effect with a presidential bid in 1965. Despite assembling a reformist and modernising programme, <u>Un Nouvel Horizon</u> (Villeneuve and de Virieu, 1981), including a mild commitment to decentralisation, Defferre was unable to sustain his bid and gave way to Francois Mitterrand. Mitterrand was a politician without a party, his political base being in the CIR club (<u>Convention des Institutions Républicaines</u>) and, as such, was able to attract general support on the left, scoring 45% of the vote in the second ballot against de Gaulle. Thereafter, he formed a federation of the left, the FGDS (<u>Fédération de la Gauche Démocratique et Socialiste</u>), bringing together the SFIO, the CIR and left Radicals and striking an alliance with the Communists for the 1967 legislative elections.

It was the 'events' of May 1968 and de Gaulle's referendum the following year, however, which were to ring the death knell for the SFIO. The established parties of the left were caught completely unprepared for the student explosion and industrial unrest and could only dither while Pompidou organised the counter-attack of the 1968 legislative elections. One of the first casualties was the FGDS. De Gaulle's 1969 referendum on regionalisation and the reform of the Senate caught the SFIO even more at a loss. Rooted in the <u>notable</u> system, it had to oppose the reforms, particularly the effective abolition of its major power base in the Senate, but without any clear alternative of its own (Meny, 1974). In the ensuing presidential election, Defferre, as the candidate of the SFIO, polled a derisory 5% and the left failed to make it to the second ballot.

Moves to re-establish the party were started almost immediately with a change of name - to <u>Parti Socialiste</u> - and a new General Secretary, Alain Savary to replace the discredited Guy Mollet. The real foundation of the PS, however, dates from the Epiny Congress of 1971 which saw the adhesion to the new party of Mitterrand and his installation as First Secretary. Along with Mitterrand's CIR, there joined a variety of left wing groupings and various independents. Within a year, a Common Programme had been signed with the Communists and in the presidential elections of 1974 Mitterrand ran within one percent of the victorious Giscard D'Estaing.

The 1970s saw a transformation of the organisation, membership and fortunes of the Socialist Party. Membership, which had doubled by the mid-1970s (Bell and Shaw, 1983) is strikingly middle-class; not the petite bourgeoisie and primary teachers who had dominated the old SFIO but the new middle class of the tertiary sector, public servants and secondary and university teachers. In the higher reaches of the party, teachers overwhelm all other occupations (Duhamel, 1982), accounting for no less than 137 of the 285 Socialist deputies elected in 1981 (Villeneuve and de Virieu, 1981). In contrast to the Communist Party, there are few manual workers and there is no formal trade union link. According to Portelli (1980) this makes the PS not so much a typical European social democratic or labour party, as an advanced radical group. There is a great deal to be said for this judgement, which helps to explain the party's openness to wider concerns than those of some of its sister parties in the Socialist International and, in particular, its ability to accommodate pluralist, decentralist and autogestionnaire ideas often shunned by the solidaristic and centralist tendencies of trade union-based parties. The expression 'pluralism', directed against an all-encompassing state as well as monopoly capital, recurs constantly in the Socialist vocabulary. Indeed, it might be tempting to put the PS down as a catch-all party of reform were it not for its insistence on its socialist mission. Despite its decidely unproletarian membership and its broad political agenda, the party's rhetoric is redolent with calls to class warfare and revolutionary promise.

The confusion is partly attributable to the inherited passion for revolutionary sloganising which the old SFIO used to combine with the most timid reformism in practice. Partly, it stems from the differing perspectives of the various party factions. The left-wing CERES group is largely Marxist in inspiration, believing in the class struggle and a powerful state and with little time for autogestion or decentralisation. For a large section of the party, the circle is squared by the 'electoralist tendency for the already vague concept of class... to be broadened yet further to include the middle classes and small medium sized businesses' (Johnson, 1981, p. 159) seem as victims of capitalist concentration. By 1974, in fact, Mitterrand's front de classe was being defined negatively to exclude only the exploiters (Johnson, 1981).

Other groups, which have joined the party since its foundation, find it easier to dispense with the

fundamentalist rhetoric. In 1974, a section of the PSU
(Parti Socialiste Unifié), an ultra-left libertarian group
which had supported the student revolt, came in, under the
leadership of Michel Rocard. The PSU had developed a strong
commitment of self-management at the workplace and self-
government in the communities and regions of France and in
1966 at Grenoble had worked out a theory of 'internal
colonialism' (see chapter three) as an ideological
underpinning (Rocard, 1967). Despite his move from the
gauchiste PSU to the right wing of the Socialist Party,
Rocard maintained his commitment to regional government as a
means for providing more efficient government, better
planning and greater democracy and for satisfying the
cultural aspirations of the regions of France.

In due course, the new party provided a home for a broad
coalition of progressive forces effectively excluded from
the old SFIO-autogestionnaires, feminists, ecologists and,
as we have seen (chapter three) regionalists, as the
regionalist movement swung to the left. These provided the
climate in which decentralist ideas could develop, despite
the continuing presence of the remaining SFIO notables.
Another important influence was the CFDT trade union and its
leader, Edmond Maire, an associate of Rocard. From the late
1960s, the CFDT had been developing autogestionnaire ideas,
linking these with the need for decentralisation of the
state. It became a strong supporter of regionalism, urging
the need for units of government in which people could take
charge of their own destinies and, in the words of the
fashionable slogan, vivre, travailler et decider au Pays
(Escafit, 1981). Although proposals for the CFDT to
affiliate formally to the PS fell through, it remained close
to the party and influenced its thinking.

THE MARCH THROUGH THE INSTITUTIONS

As early as Defferre's programme of 1965, we have seen the
Socialists were reconsidering their traditional centralist
attitudes. The realignment of the left and the advance of
the new party were to bring the issue to the fore and, in
particular, to focus attention on two questions, the
dismantling of state controls on communes and départements
and the development of new regional institutions. We have
noted the transformation of the regions of France and the
emergence of social and political movements seeking regional
self-government. New, modernising elites - the famous
forces vives - repelled by the notable-ridden SFIO, found a
more congenial home in the new PS. As early as 1965,
Michel Phlipponneau, the Breton regionalist and CELIB

activist, had decided that regional government could only come through concerting the parties of the left and many others followed his example. Support for regional government also came from modernisers within the party who saw the inadequacy of the existing local government structures for their visions of economic and physical planning. Regional government could, it was hoped, both provide the institutions for radical policy initiatives and encourage the emergence of a new type of politician, more innovative, less tied to the notable system and more policy-minded.

The commitment to decentralisation was immensely strengthened by the Socialist advance in local government in the course of the 1970s and the emergence of prominence of younger, more dynamic municipal leaders. The absence of a strong commitment by the Socialists to the Gaullist constitution and the lack of alternation at national level for the first twenty three years of the Fifth Republic concentrated attention at local level where the party began a 'march through the institutions', building on local election successes to the national triumph of 1981. Of the towns of over 30,000 population, Socialist (SFIO, then PS) mayors won 32 in 1965, 40 in 1971 and no less than 80 in 1977. The Communist tally went up from 34 in 1965, to 45 in 1971 to 72 in 1977, when it was in alliance with the PS, presenting joint lists in most cases. Including other left wing forces such as the MRG (left Radicals) 153 out of 221 such towns were controlled by the left after 1977 (Frears, 1977). The 1976 cantonal elections (for the conseils généraux of the départements) had showed the same picture, with strong advances for the left. Despite the bias of the cantonal election system - with gross overrepresentation of the rural areas - the left now controlled 41 out of 96 departments, 14 more than in 1973 (Frears, 1977). Following these victories, the left automatically advanced in the indirectly-elected regional councils, the urban communities and the Senate.

One result of this advance was to bring to the fore younger, more dynamic municipal leaders who saw their role as the advancement of socialism through local action. A forerunner of this process was Grenoble, for long governed by traditional notables of right or left, which experienced a boom in the 1960s in advanced, science-based industries. The middle-class incomers who arrived in its wake soon grew impatient of the old politics and the inability of the municipality to expand and maintain services at the level of their expectations. One of the newcomers, Hubert Dubedout, formed an independent list of candidates which, allying with the Socialists, defeated the right-wing municipal council at the 1965 elections. In 1967-8, Dubedout helped Pierre Mender-France in his brief political come-back and in due course joined the <u>Parti Socialiste</u> and became a Deputy (Ardagh, 1982). The city's environment and public services were transformed, with a new airport, roads and, more recently, a telematic information system extending into all the <u>quartiers</u> of the town. The city was divided into six sectors, in each of which community councils debated local priorities and made proposals to the council. Socialist mayors elected in the 1970s sought to follow this example, expanding public services and seeking to open up the process of local decision-making.

The range of innovative activity was immense. Many towns took up the theme of energy conservation, or the use of solar power; in Carcassonne, the whole public lighting system was run on solar energy. Many towns expanded their cultural provision or extended it in a more popular direction. In Besancon and other cities, social security payments were increased to the old, large families and single parents. All over France, public transport was expanded and town centres pedestrianised. Ecology was taken furthest in La Rochelle, where the MRG (left Radical) mayor, Michel Crépeau, halted the building of skyscrapers and, for a time, even provided free municipal bicycles. Of course, it was not only left-wing councils which innovated or spent freely – subsidised public transport was a largely non-partisan issue and Jacques Chirac's Paris has a notably large social security budget – but the effect on the Socialist Party of the new activist local politics was considerable.

One feature that did distinguish left-wing councils was a growing economic interventionism, which we trace in more detail in chapter seven. Both the traditional councils of towns and <u>départements</u> and the new regional councils set up in 1972 were used to this effect. In 1979, a major clash occurred in Marseille when the prefect vetoed Defferre's attempt to use the city's resources to save the shipbuilding group, Terrin. In the solidly socialist Nord Pas-de-Calais region, a substantial <u>cabinet</u> was built up by the President, Pierre Mauroy, despite the executive power being formally vested in the prefect. Economic plans were developed and major initiatives taken in transport, infrastructure and industrial development. Here and in other socialist regions, the powers of the 1972 Act were pushed to their limit and sometimes, with prefectoral connivance, beyond it.

This experience, and the battles with a central government determined to control it, undoubtedly served to strengthen support for the regional idea within the Socialist Party (Chevallier, 1982).

More generally, the reliance on local government as the party's main power base and the vigorous use of it by the more dynamic municipal and regional leaders was absorbed ideologically by the Socialists, producing a philosophy of local power and showing the viability of local routes to socialism. As we have seen, decentralisation had been mooted as part of the new socialism as early as Defferre's Nouvel Horizon of 1965. In the Common Programme of 1972 and Mitterrand's presidential platform of 1974, decentralisation was promised and rather imprecise measures of regional government proposed.

At its 1971 congress at Avignon, the PS adopted a regionalist resolution, confirmed when the Convention of Suresnes included the commitment in the programme of the new party (Mény, 1974). The commitment was refined and strengthened in the course of the 1970s, receiving its fullest expression in the 1981 document La France au Pluriel (Parti-Socialiste, 1981). This, in a remarkable revision of traditional thinking on the French left, recognised the need for centralisation under the Third Republic to protect the revolution and extend democracy, human rights and public services but condemned the increasing centralisation of the twentieth century which it blamed on the natural tendencies of capitalism. Regional cultures and languages had been suppressed and swamped in the same way as French was being swamped by the English language and American culture. Centralised regional policies unbalanced local economies. Local businesses had gained little from the tourist developments on the Mediterranean; local cultures had been devalued to tourist spectacles and holiday homes had pushed up land prices. The party espoused the solgan Volem viure al pais but this required that people should have work. The Socialists believed that this meant industrial work because of the fragility of the teritary economy and because a region cannot live by management and services alone. The document saw three possible models of regional development policy:

1. Providing incentives to private industry as in the past. This had been shown not to work;
2. Nationalisation and management of the regions from the centre as in the USSR. This would not work either and would inevitably lead to policies favouring the centre as against

the regions;
3. Increasing the power and resources of the regions themselves to do the task. This was the solution favoured by the party. So decentralisation would be underpined by an ideology of indigenous economic development, the key level of which would be the region. Building up regional economic power would attack not only state power but also that of capitalism which was itself centralised and in tight collusion with the central state. So socialism and decentralisation, far from being incompatible, were part of the same struggle. This is not quite the full internal colonialist argument but it is close enough to it to strike a chord with the micronationalists. The implication is that a regional economic sector, based on small and medium firms and decentralised public enterprise could be developed, with local political power being used to protect it against centralised big capitalism.

Specifically, the Socialists proposed the direct election of the regional councils, whose boundaries would be redrawn to reflect more accurately regional loyalties. There would be a reform of finance to increase the resources available to the regions. Regional investment banks would be established and the national economic plan reformed to allow more regional influence. Regional languages would be encouraged in schools. The programme for the 1981 elections also promised a special statute for Corsica and a separate département for the Basque country. It was a programme of cultural, economic and political measures which would satisfy the demands of most regional activists, few of whom were seriously pushing for separatism or even federalism.

Regionalism, however, was but one dimension of the more general problem of decentralisation and whatever pledges the party made in that respect would have to be reconciled with its promises on the communes and départements. These were also worked out in the late 1970s, against the background of the Guichard Report (See chapter two) and the legislative proposals of the Giscard government.

The Parti-Socialiste was, from the start, highly critical of the Giscardian proposals, seeing in them an attempt to undermine left-wing strength in he localities through the creation of a new level and the eventual merger of départements and regions - though it seems that some of the big-city mayors appreciated that their influence might be enhanced. Detailed criticisms focussed on the weakness of the proposals for the regions, the reliance on indirect election and finance. At a more philosophical level, the whole scheme was condemned as 'technocratic' and 'dualist'.

'Technocracy' and 'dualism', concepts of some importance to the intellectuals of the <u>Parti Socialiste</u>, merit a word of explanation. 'Technocracy' refers to the subordination of political choices to technically-defined and justified imperatives; and to the remoulding of state institutions around technically-defined activities rather than politically-defined communities. Socialists have attacked <u>ad hoc</u>, unelected agencies, the power of the <u>grands corps</u> and the technical <u>tutelles</u> which these exercise over local government and which are recognised as increasingly more onerous and damaging than the largely redundant prefectoral <u>tutelle</u>. In a wider sense, the attacks on technocracy have been aimed at much of the reasoning which underlay the search for 'rationality' and 'effiency' idn the reform of British local government and which has recently come under attack for its weak intellectual base (Dearlove, 1979).

'Dualism', a concept owing not a little to the French intellectual fondness for Cartesian abstractions, was taken up in the <u>Parti Socialiste</u> as a criticism of Giscardian liberalism. Under Giscard, the state was seen as surrendering its relative autonomy to multinational capital to produce a two-speed society. On the one hand, there would be the modern sector, competitive and open to the world market, a 'space for managers and supersonic technocrats' (Charzat, 1981); on the other, the protected sector, 'archiac but more convivial', onto which the social costs of the first sector would be unloaded. Giscardian decentralisation was seen as a means of furthering this by separating state and local matters, retaining economic power at the centre (in collusion with capital) but thrusting the social costs of adjustment onto the periphery. The liberty which was being soughts was not the liberty of the localities but that of central government, which would have an enhanced freedom of manoeuvre behind the protective screen of local representatives. According to Worms (1980) a whole dualist society was thus being created of economic sector-social sector; competitive sector-protected sector; State-local government. A socialist strategy for decentralisation would have to avoid this trap, notably by giving the localities a hand on the levers of economic power, allowing communities to take in hand their own development and challenging the logic of the corporate market.

The main problem confronting the <u>Parti Socialiste</u> in devising its own strategy, however, was the fact that, while there was wide support for decentralisation in principle, in practice it meant different things to different elements within the party. There were, as we have seen, regionalists who looked for a radical break with the traditional structure of French local government. Even these were divided into modernisers who saw the region as a planning unit; micronationalists from the periphery; and a few exponents of the old PSU's internal colonialism thesis. These competed for influence with the exponents of municipal power and the notables from the départements. Big city bosses, who already enjoyed a considerable measure of independent power, were often suspicious of regional cnetures. The problem of the small communes was quite intractable. It was easy to condemn proposals for forcible mergers as technocratic engineering, digging up the vital roots of French democracy, but unless communes were enlarged it was absurd to talk of increasing their powers and resources - the technocratic logic does have some force. Yet, while almost everyone was agreed that 'something must be done' about the small communes, in practice their inviolability was as firmly entrenched in the Parti Socialiste as elsewhere.

The party's differences came to a head in 1977-8. At the 1977 conference of Socialist and Left Radical Presidents of conseils généraux, Mitterrand pledged that decentralisation would be among the first acts of a new Socialist government; but there was little agreement on how it was to be done. Edgard Pisani had produced a report calling for radical reform. He was answered by a notable of impeccable credentials, Andre Chanderganor (mayor of Mortroux since 1953, sometime president of the conseil general of Creuse and of the regional council of Limousin, member of Parliament since 1958). Chandernagor attacked the 'visionaries' and warned against leaving the state stripped of powers. Among the 'visionaries' were those whose main concern was with planning and aménagement du territoire and who tended to see the region as the best level for this. Their fear was that, if priority were given to the départements and communes, the regions could find themselves squeezed out with consequent damage to the ability of the system to plan and innovate as opposed simply to managing.

In 1978, the problem was handed to a committee under Pierre Mauroy. Although one of its early ideas, for a new tier of government at the cantonal level, with the départements indirectly elected from cantons and communes, did not find wide favour, the other proposals were accepted as the basis for the proposition de loi submitted by the

Socialist group to Parliament in 1979, in reply to the government's scheme. This in turn bore a close resemblance to the scheme proposed by the Socialist Government in 1981.

No change in the basic three-tier structure of local government was proposed, as this would merely mobilise opposition. Instead, the idea was to build on what already existed, to set in train a dynamic process of reform by freezing elected representative from their centralist shackles. All three levels would, it was hoped, have a clearly defined role, the region in planning and aménagement du territoire, which would become more important under a government of the left; the département in management of services; the commune in urban development and small-scale interventions. Tutelles would be lifted and all three levels allowed to function freely. Linked to this would be a reform of local government finance; an improvement in the status of councillors and an opening up of opportunities for election; and a reform of the local public service to create a class of local officials with their own career structure. Local pluralism would be encouraged by introducing proportional representation in regions and large towns.

There remained the problems of how to develop real economic powers, how to reconcile decentralisation with the rest of a Socialist programme involving measures such as nationalisation, planning and the promotion of social and economic equality, and how to ensure coherent policy making in a system with so many levels of government. At the ideological level, it was necessary to show how decentralisation could, as claimed, constitute part of the 'rupture with capitalism' to which the party was, in theory at least, committed. There was always the danger that decentralisation could end up by merely establishing a new 'dualism'.

The key element here was devising a means of economic intervention for local government. Localities were to be ebcouraged to develop local centres of economic decision-making as a counterweight to the Parisian Political-economic power nexus. This involves accepting the profoundly unmarxist notion that dispersed political power can counterbalance concentrated economic power. In practice what seemed to be involved was a role for local government in building up the small and medium-sized business sector, tapping energies neglected by the corporate giants of the centre. This was part of the 'rupture with capitalism' in the sense in which, we have noted, capitalism tends to be defined on the French left, as centralised big business. Locally-based small businesses were seen as part of the progressive alliance.

Decentralisation would also free the central state itself for constructive and purposive interventions, on its own and in partnership with the liberated localities The division of work among the levels of government would not be a rigid functional one, with the attendant dangers of 'dualism' and incoherence in policy making. Rather, 'in each domain and over the same territory, several interventions (would be) possible, thanks to as rebalancing of the powers (of the various levels) allowing a real negotiation between local governments representing the needs and demands of citizens and local communities and the apparatus of the central state' (Worms, 1980, p. 30). So what observers such as Ashford (1980) see as the strength of the French system, its unitary nature and the absence of functional divisions, would be retained; but the balance of local and central power would be shifted in favour of the former and many problems would be resolved at lower levels rather than always drifting towards the centre. The latter tendency would be further discouraged by a strict limitation on the cumul des mandats.

Coordination and coherence among the levels of government would be ensured by a revived system of planning which would not be centralist and dirigiste but would provide a forum for negotiation over objectives and means. It would also help to protect overriding national priorities, reconcile sectoral and territorial needs and prevent competitive overbidding by regions and localities for mobile industry. The latter was not, though, considered to be a major problem, as local intervention was seen as lying in the domain of indigenous industry and local enterprise rather than in the attraction of multinational capital. The Plan, together, with other economic and financial measures, would ensure equity in the distribution of resources, allowing poor regions the same opportunities as wealthier ones.

With horizontal coordination in government improved and accountability enhanced, intergovernmental relations would be conducted on an open, political basis, removing the obscurity which encouraged power manipulation in the old system. The repoliticisation of local affairs, begun by the new municipal and regional leadersm was a vital theme and one which implied an assault on the notable system. Of course, strengthening community government had its dangers, notably that of 'local totalitarianism', with more pressure on non-conforming individuals than might be found in a larger grouping. To combat this, the Socialists would institute proportional representation and encourage the growth of associations and the institutions of civil society. And how would the individual be protected in turn against his associations? This would, according to Worms (1980) take us 'to the very heart of the problem of autogestion'.

Decentralisation was thus directly linked into Socialist ideology. The jacobinism of the past was excused on the grounds that it was necessary in its time to secure the republic, overcome provincial reaction and ensure social equality. Now political advance required the opposite, the Project Socialiste (Parti Socialiste, 1980, pp. 252-3) declaring that 'to decentralise is to put into place one of the most powerful levers of the rupture with capitalism, which will allow the citizens to take the most direct part in the immense task of social transformation.' Leaving aside the typically extravagant rhetoric, it is clear that decentralisation was being seen as an integral part of the strategy and not as something which cut across it. Serious problems, however, remained. Socialist advances in local elections helped the policy gain acceptance by presenting the prospect of united Socialist central and local governments moving in the same direction but there were still strong jacobin elements in the party, including, it was often suspected, Mitterrand himself. Certainly, as President of the conseil général of Nièvre, he was firmly attached to the institution of the département, so that his defeat of Rocard's 1980 challenge for the presidential nomination was a setback for the regionalist cause. Crucial ambiguities remained about the Socialists' commitment to an attack on the notable system. The refusal to take on a reform of the three-tier structure of local government but to devolve power to the existing units entailed a reinforcement of the old elites. On the other hand, direct election of regional councils could, if it were enacted soon enough, set in train a dynamic which would progressively erode the powers of the départements.

Finally, there was the problem of conflict with other items in the party programme, notably nationalisation and strengthened national economic planning. If real economic powers were to be given to local councils, these would have to be accommodated within national policy priorities. In order to reconcile national and local priorities, a great deal of reliance was to be placed on the revival of planning and the contractual principle whereby the State and the localities would bind themselves to negotiated collaborative programmes. As we shall wee, reviving planning in a period of uncertainty and scarce resources was to prove a major problem in itself.

By the Presidential election of 1981, then, decentralisation was firmly established on the Socialist agenda. The 110 propositions which formed Mitterand's campaign platform - effectively pushing aside the Project Socialiste - including pledges on the direct election of regional councils; the transfer of executive authority in regions and départements to the presidents of the councils; the end of the prefectoral tutelle; a reform of local finance; new responsibilities for local government; the promotion of regional languages; a new département for the Basque country; and a special statute for Corsica. In May 1981, the 110 propositions became the Socialist Manifesto, with a clear promise of early action in the event of victory.

5 The programme in Government

After the election victory of 1981, Gaston Defferre, mayor of Marseille, was appointed Minister of the Interior and Decentralisation and swiftly brought forward the first proposals for action. Since then, there has been a constant stream of legislation, decrees and circulars implementing the programme. These have had three main impulses (Dupuy anfd Thoenig, 1983); firstly, to shift power relationships 'horizontally' in favour of elected politicians, against administrators; secondly, to transfer functions vertically downwards to local councils; thirdly, to give the councils the financial and other means to fulfill their new responsibilities. As agreed in opposition, no attack was to be made on the basic three-tier structure of local government. This was largely a matter of political prudence, as even if the Government had been able to agree on a new structure, it would merely have mobilised opposition in Parliament and in the localities. Ministers were able to add in justification of this procedure that theirs was a political reform, aiming to give power to the elected representatives of the localities, not ti invent new technocratic forms to fit the convenience of the centre.

SHIFTING THE POWER BALANCE

The first law, that of March 1982 on the Rights and Liberties of Communes, Départements and Regions, was concerned almost entirely with the first objective, shifting power relationships. The idea was to give an abrupt shock to the system, break existing power patterns and set in train a decentralist dynamic. The prefectoral tutelle over the communes was formally abolished. Following a ruling in the Constitutional Court and a consequent amendment of the law, communal acts now became legally effective from the moment they are notified to the 'representative of the state' (i.e. the Prefect). He can, if he thinks they violate the law, refer them to the Council of State but otherwise has no power over them. So the administrative tutelle is replaced by a post-facto legal control. At the same time, technical and financial tutelles were to be reduced and codified in legal form. Mayors and Presidents of councils are given the right to order payments from the Trésoriers Payeurs Généraux, subject only to control of their legality in a series of new regional courts of accounts.

In the départements and regions, the prefect ceased to be the executive authority and was replaced in this role by the President of the council. Regional councils were to be directly elected and, at that time, would become fully-fledged collectivités territoriales (local authorities). Meantime, they would retain the more modest status of établissements publics. For his part, the prefect, renamed commissaire de la république, gained the power to direct - and not merely coordinate - the field services of the various ministries, excluding defence, education and justice. In a 'functional clause, whose presence in such a general bill owed a great deal to Defferre's experience in Marseille, it was explicitly laid down that regions, départements and communes were to have the right to help firms in difficulty.

In Parliament, the Right made a strong show of opposition to the proposals. Michel Debre, de Gaulle's former Prime Minister and the most obdurate of jacobins, described them as a threat to the unity of the state and insisted that direct election of regional councils was an expression of 'regional sovereignty' in contradiction to national sovereignty. Olivier Guichard, author of the 1976 report on decentralisation, Vivre Ensemble, found the 'unity of the Republic threatened.' The Senate, dominated by opposition notables, tried to substitute a complete text of its own. With its massive majority in the Assembly and public opinion favourable, however, the Government was able to prevail.

Once the initial excitement about the changes had died

down, it was possible to make some assessment of their significance. in many respects, they have simply legitimised what was already happening. The prefectoral tutelle had long ceased to be onerous and powerful presidents in the départements and regions had already gained a share of the executive power. In the large cities, mayors already held sway. In the small communes, the new freedoms would be an empty shell, given the lack of the means to exercise them. Too cynical an interpretation of the changes, however, would not be in order. The effects, especially at the departmental level, have been considerable.

The law laid down that a convention must be negotiated in each département dividing the services of the prefecture between the President and the commissaire de la république. Where staff were not transferred, staff were to be 'placed at the disposal' of the council as needed. Generally speaking, councils controlled by the Right initially allowed the commissaire de la république to retain the bulk of the staff, while those controlled by the left sought greater transfers. In several cases, councils have engaged the services of members of the prefectoral corps (who retain the right to come back into central government service). By 1983 there were some seventy of such secondments and, although these were frequently right-wing prefects uneasy about working for a left-wing government, such moves do indicate an awareness that power does now revolve around the President of the conseil général and the city mayor. Most Presidents have in most cases built up a substantial staff of their own, in some cases to such a degree that the Ministry of the Interior, worried about the cost of a burgeoning bureaucracy, issued a circular warning against it. There has also been a great deal of argument about prestige property such as premises and cars, as the Presidents have sought to project themselves as the new leading citizens of their départements.

The rise of the département has been most noticeable in rural areas where the communes, unable to make a reality of their new freedom, are forced to depend either on the commissaire de la république and the external services of the state or on the département's help. Départements are allowed to establish technical services to help the small communes and, although the law lays down that no local authority shall exercise a tutelle over another, relationships of dependence inevitably arise. In many areas, the President of the conseil général has taken on the role of an urban mayor (Dupuy and Thoenig, 1983) as the political 'boss' of his territory, with individual councillors exercising similar sway in their own cantons. In other cases, mayors have sought to cling to the commissaire de la république and his adjoints (the old sub-prefects) as representing a less partisan source of support. Indeed, the loss of the protective 'screen' of the prefectoral tutelle - mayors must now face the responsibility for their actions alone - and the fear of a more irksome and political departmental one left many mayors in two minds about the merits of decentralisation. The 1981 Congress of the Association of Mayors failed to agree on a resolution welcoming decentralisation precisely because of this (Information Municipale, Dec. 1981). Some observers (Dupuy and Thoenig, 1984), indeed, claim that the removal of the tutelle had nothing to do with a desire to free the mayors but was intended simply to increase their political accountability. The weakness of so many communes means that the commissaires de la république and their adjoints retain an important role in advising and mediating among local authorities despite the loss of their executive and tutellary powers. The presence of so many members of the prefectoral corps in the service of local councils but retaining close links with their state colleagues has made this a great deal easier. There is even an association for 'senior state civil servants in service with local authorities and regions'.

The Commissaire de la république was, as we have noted, to have his power over the external services of the various activities strengthened. These, in turn, were to deconcentrate their activities to field offices so that decisions could be taken on the ground. The idea was to accompany decentralisation with deconcentration, to give local councils a powerful and single interlocutor at local level and so prevent cases drifting up to Paris. Despite the constant exhortations of the Ministry of the Interior, this does not appear to have happened. In March 1983, a meeting of the Comité Interministériel de l'administration territoriale was resolving to improve matters but, by mid-1984. Defferre admitted that he was still having to tell his ministerial colleagues not to go into the localities to sign conventions, as henceforth the signature of the commissaire de la république alone committed the state (Les Nouvelles, 23 May, 1984). Because ministries have not deconcentrated decision-making to their field officials the commissaires de la république have not asserted effective control of the latter. Certainly, all mail now passes through the prefecture but control of telephone communications is another matter. In another evasion of prepfectoral control, all departmental directors of équipment now meet their Parisian superiors together once a month (Dupuy and Thoenig, 1983). Codification and scrutiny of technical tutelles is to be undertaken by a special commission but some observers have doubted whether, given the complexity and rapid changes in technology and administration, it will be possible to limit technical tutelles by such a formalised device. Generally speaking, the officials of the state technical services retain great scope for playing off the commissaire de la république off against the local authorities and the latter off against each other.

The consolidation of the département has been so marked as to worry many regionalists, fearful that, by the time direct elections - repeatedly postponed - come, the département will have occupied the available political space. Regional presidents, it is true, tend to be figures of national stature, which is not always true at the departmental level, and they have made the most of their new powers; but lacking the legitimacy of direct election and limited in their functions (see below) they are at a serious disadvantage. As is the case for so much of the reform programme, experience differs from one part of the country to another. In those regions where an energetic and purposeful policy had already been pursued - by concentrating the limited resources on priority programmes, for example - the region has been able further to consolidate its position. In other regions, policies of saupoudrage continue, with moneys scattered evenly across programmes and localities. In these cases the region, lacking its own programmes, is unable to exercise much leverage on those of other actors.

Two vital Socialist proposals for strengthening political control involved the limitation on the cumul des mandats and measures to improve the availability, training and remuneration of elected members. The cumul was criticised for obscuring accountability, causing issues to drift to the centre and encouraging absenteeism. The Parti Socialiste had for some time operated an internal rule limiting the cumul to one local and one national office, though this seems to have been honoured in the breach as much as the observance - in 1983 a third of Socialists deputies had at least three mandates. In 1982, a report commissioned from the Socialist Senator Marcel Debarge proposed limiting the cumul, increasing the allowances for members, allowing for full-time service in key offices and statutory rights to time off work for elected officials. After considerable discussion, the Government produced a bill in 1983 to implement, in as economical a form as possible, all the major recommendations except for that on the cumul des mandats. The vested interests in the party against limitation had been strengthened since 1981, with the legislative victory giving control over more regional councils and several ministers moving to consolidate their power bases by taking the presidencies. By 1983 there was a further fear of creating by-elections by relinquishing mandates. Essentially, though, the problem was the reluctance of the political class as a whole to give up any of its traditional advantages. Government and Opposition quietly colluded to sweep the issue under the carpet. Instead, excessive cumulation is to be discouraged by limiting the total indemnities from all offices to one and a half times the Parliamentary salary - a deterrent only to the most mercenary of politicians!

Shifts of power have also taken place within local councils. The new rights and responsibilities and, as we shall see, the new functions, have tended to be vested in mayors and presidents of departmental and regional councils rather than in the councils themselves. This has reinforced the existing tendency to executive dominance, with the mayor or President choosing his adjoints (or deputies, in charge of specific services or functions) and, in many cases, effectively nominating the councillors themselves. In some départements, mainly those controlled by the Socialists, the opposition is given seats on the bureau, which meets regularly between the infrequent meetings of the conseil général; this responds to a complaint of the left from their opposition years, when they tended to be excluded. Executive power, however, remains in the hands of the President, who may organise his adjoints into an executive committee. With the failure to tackle the cumul des mandats and the rise of departmental bosses alongside those of the cities, this is leading to a formidable strengthening of the existing political class. Radicals and autogestionnaires have been quick to point this out (Eme, 1984), claiming that the promises of a new citizenship have been betrayed.

The only major changes involving local democracy have been the change in the electoral system and the so-called 'PLM law' for Paris, Lyons and Marseille, the latter owing as much to partisan self-interest as to an aspiration for greater democracy. Under the Fifth Republic, the local electoral system had been changed several times, mainly for partisan reasons, and by 1981 was highly non-proportional. For the conseils généraux of the départements, there was one councillor per canton, elected on a two ballot system similar to that used for parliamentary elections. The main complaint about this was not that the cantonal system gave non-proportional results, because, especially in the rural areas, it is considered vital for a councillor to have close links with a constituency, but the maldistribution of seats. Wright (1978) cites the examples of the cantons of Embrun and Barcelonette in the Hautes Alpes which, with populations of 7191 and 318 respectively, each had one representative in the conseil général in 1976 and of the Haute Garonne where in 1973 Toulouse, with a population of 450,000 had four councillors, while the rest of the département, with a population of only 250,000, shared thirty-five councillors. In general, there was a pronounced bias towards the rural areas and, consequently, against the left. For the 1982 elections, a fairly hasty redistribution was undertaken, though large discrepancies remain and nothing has been done to shake the grip of the departmental political bosses.

In the communes, the system was more complicated. The number of councillors was determined according to the population of the commune and there was everywhere a two ballot system of voting for lists of candidates. In the communes of less than 30,000 population, the elector could vote for a whole list or for candidates from more than one list. Lists could be merged or modified between ballots, so allowing bargaining and coalition-forming following the first round. In the two hundred or so towns of over 30,000, cross-voting between lists was disallowed, as was modification of lists between ballots. The list gaining the absolute majority of votes on the first ballots or a plurality on the second took all the seats, giving no representation to the opposition. The five large cities of Paris, Marseille, Lyon, Toulouse and Nice had the same blocked list system but operated on the basis of sectors, each of which elected a separate list; so the opposition, by winning in individual sectors, could be represented on the council (Frears, 1977).

This system had been strongly criticised by the left who, no doubt rightly, saw it as a means of excluding them from power in many towns and cities. By the 1970s, however, it had begun to work in their favour. Firstly, it forced unity on the left, who could not afford to split the second ballot vote; accordingly, they had either to forge coalitions before the first ballot or else withdraw the less popular left list altogether at the second. Following the Union de la Gauche in 1972, Socialists were obliged to take Communists onto their lists and sever local alliances with centrists. Secondly, the system exaggerated the left's success in the municipal elections of 1977, when they won 159 of the 221 cities of over 30,000 population.

Objections in principle to a system which allowed local bosses to keep such a tight grip on the towns and stifled opposition between elections, however, remained and the Parti Socialiste was committed, albeit vaguely, to proportional representation. So the search began for a system which, without damaging the left's prospects, could introduce more proportionality, while at the same time allowing strong party government. After months of speculation and debate about the partisan implications of various schemes, a system was found for the 1983 municipals. In towns of over 3,500, there is a list system with two ballots. if a list gains the absolute majority at the first ballot, it is allocated half the seats, with the remainder distributed proportionately among all the lists, including the victorious one. If there is no list with an absolute majority at the first ballot, then a second ballot is held. The list gaining a plurality (i.e. the largest single number of votes) at the second ballot gains half the seats, with the rest divided proportionately among all lists including the winners. There is no cross-voting between lists and no alteration or merging of lists between ballots. So the mayor at the head of his winning lists is still guaranteed control of the council while at the same time the opposition is represented.

The representation of the opposition could help to open up local political life. Not only parties but effectively whole areas of towns were effectively disenfranchised in the past because of their support for losing lists. Now opposition representatives will have a continuing political platform and, by pursuing local issues and complaints, the means to build a political base. This could be particularly true of the poorer quarters of those cities traditionally controlled by the right.

The other major electoral change, which eventually produced the PLM (Paris, Lyons, Marseille) law, was from the start surrounded by partisan rancour. In 1982, Defferre proposed that the councils of the arrondissements of Paris which, since 1975, had had advisory powers, should become fully-fledged councils in their own right, leaving the mayor and city council with greatly reduced powers and weak coordinating role. The idea had been floated some time before by the Socialist councillots in Paris but, in the circumstances, it was bound to be seen as a crude attack on the power base of Jaques Chirac, mayor of Paris and national leader of the Gaullist RPR opposition. Such an impression was not dispelled by the fact that, while the proposal was defended as a means of decentralising power and involving the electors more closely in municipal affairs, the other big cities, including Defferre's own Marseille, were not included. In the ensuing battle, the Government was forced to back down, making two major concessions. The powers of the new mayors of the <u>arrondissements</u> were much reduced, leaving Chirac with his power largely intact; and the scheme was extended to Marseille and Lyon.

Each of the three cities now has <u>conseils d'arrondissement</u> and a <u>conseil municipal</u>, elected at the same time from constituencies known as <u>secteurs</u>. In each <u>secteur</u>, lists are presented and the elections conducted on the same basis as for the larger communes. Where a <u>secteur</u> comprises a single <u>arrondissement</u>, the candidates at the top of the lists will take the municipal council seats in the appropriate proportions. Those coming further down the lists will be given seats in the <u>conseils d'arrondissement</u> (in which the municipal councillors also sit by right) until all seats are filled. Where a <u>secteur</u> comprises more than one <u>arrondissement</u>, successful candidates can choose, in order of their position on the list, in whch <u>conseil d'arrondissement</u> they will sit.

The <u>conseils d'arrondissement</u> will have no taxing powers but will look after creches, youth facilitis, gymnasiums, public baths, gardens and open spaces and old people's homes. They will also carry out some state functions previously performed by city mayors, such as registration of births and deaths and compiling lists for entry to schools and military service. Their most important role, however, is now seen less in terms of administration on their own account than as sounding-boards for local opinion, notably on planning applications and redevelopment schemes. While there are fears that this will slow down decision-making, it does represent one of the few attempts to pursue decentralisaiton within local government itself.

The outcome of the PLM affair was a considerable victory for Chirac in his improbable new role as defender of local democracy. In the 1983 municipal elections, he not only retained the mayoralty of Paris but, winning all the <u>conseils d'arrondissement</u>, announced that he would 'apply the law, but in a restrictive manner' (<u>Le Monde</u>, 31 March 1983). Elsewhere, the 1983 municipals saw an ebbing of the socialist tide which, coming on top of reversals in the 1982 cantonal elections, was to have considerable political significance. After setbacks in the first round, they recovered at the second ballot and list just half the gains they had made in 1977. Marseille was narrowly retained by Defferre (who had introduced some judicious boundary changes to help his own chances), though the opposition won two of the conseils d'arrondissement. In Lyon, Francisque Collumb's UDF-RPR ticket triumphed comfortably. In all the left lost 31 of the big towns, leaving them in control of 128 of the 221.

By the standard of British mid-term local elections, this was quite a respectable result. Yet the psychological impact of the reversal of what had seemed an unstoppable socialist ascent was shattering. The opposition seriously tried to claim that the government had lost its legitimacy and should go to the country. The Socialists themselves were bewildered and confused. Hardly anyone made the point that the real political test of the decentralisation strategy would be whether the central government could lose control of local councils to the opposition without peaceful coexistence between the two levels breaking down. The principles of alternation in government and dispersal of power - neither hitherto strongly rooted in French political culture - were on trial. In the event, the controversy died down and the new right-wing mayors proceeded to make the most of their new powers, to such a degree that by 1984 the burden of the opposition's complaints was not the decentralisation programme itself but the delays in implementing it. On the government side, there was certainly more dragging of feet after the 1982 cantonals and the 1983 municipal elections and, as the Socialists came under increasing political pressure, a strengthening of the jacobin wing. This was one of the causes of delay in proceeding to the next phase of the programme, the transfer of new functions to local government.

NEW FUNCTIONS FOR LOCAL GOVERNMENT

The second main theme of the decentralisation programme was the transfer of functions down from central to local government. This proved to be an extremely complex matter. French local councils have traditionally enjoyed a 'general competence' to do anything within the law in the interests of their communities. As we have seen, this was in practice tightly constrained by administrative, legal and financial controls, but the principle of general competence was closely tied in with that of the unity of the state. As part of the state, local councils had a general responsibility for public welfare but one to be exercised as part of the unitary political system. If political and administrative controls were to be lifted from local government, it would be necessary to specify their functions more precisely. This the government initially intended to do, largely replacing the general competence with a power of 'attribution', giving each level of government specific tasks to perform, and using the lists of functions already prepared under the Giscard government. In general terms, the communes were to be responsible for urban development and planning, the <u>départements</u> for personal social services and redistributive functions and the regions for planning and economic intervention. Defining more precisely and then devolving functions, however, proved more difficult than anticipated. Some of Defferre's ministerial colleagues proved less than enthusiastic. According to the press comment (<u>Le Point</u>, 26 June, 1982), the laurels for a positive attitude went to Edith Cresson at Agriculture, with Marcel Rigout (Professional Training) and Nicole Questiaux (Social Security) also cooperative. At the other extreme was Alain Savary, Minister of Education, traditionally a highly centralised service. Supported by the powerful teachers' union, Savary successfully opposed any transfer involving the content of education or the employment of teachers; he also kept his regional officials out of the extended powers of the <u>commissaires de la république</u>. Roger Quillot (Housing) opposed any decentralisation of his work on the ground that housing was a 'national priority', provoking the retort that this presumably meant that decentralisation was not a national priority. Most vigorously opposed of all was Jack Lang (Culture) who argued that as, hitherto, there had been little national cultural policy, it would be a mistake to dismantle what was just being constructed. With the President's support, he urged that culture needed to be promoted vigorously through a national policy in those regions in which it had been neglected.

An early draft bill of over four hundred clauses seeking to specify functions raised so many objections that, several drafts later, it was decided to settle for a law laying down the general division of responsibilities, with transfers to be made by order over a period of years. As far as possible, blocks of functions were to be transferred, but these were not to be defined exclusively or rigidly allocated to one level or another. Rather, the principle of shared functions would be widely applied, with each level having the lead responsibility in certain areas but acting in concert with the others. Even this proved such a task that the first set of proposals sent to the Senate in October 1982 contained only some of the transfers, the rest being dealt with in a 1983 bill. Functions were to be transferred in stages up to January 1986.

Communes are given increased powers over planning, including the long-sought right to grant planning permission (permis de construire) but subject to plans at the communal and intercommunal level and to procedures safeguarding departmental, regional and national interests. The new procedures are less a drastic upheaval than a reversal of initiative and leadership roles. Previously, the local plan (Plan d'Occupation des Sols) was produced by a working group convened by the prefect and including representatives of local councils and central ministers. Planning permissions (permis de construire) were often in fact given by the mayor, acting on behalf of central government. Now the mayor takes the initiative in preparing the plan and sends it to the commissaire de la république who consults the relevant central departments. In the case of objections, complex administrative and judicial procedures are available; otherwise, after a public enquiry, the commune publishes the plan itself. If the commune does not have a local land-use plan, then the right to grant the permis de construire remains with the commissaire de la république; and so fearful were some mayors of their new responsibility that a Senate amendment provided that, after each election, any mayor could hand the power back to the commissaire de la république. The latter in any case retains the power in respect of State, regional or departmental works and those of foreign governments and international organisations.

Départements receive new responsibilities for personal social services, certain social benefits and public health. The state remains responsible for the main social security and unemployment benefits. For those benefits which they administer, départements are subject to national minimum payment levels, though they may increase them at their discretion. Départements also take over responsibility for school transport, an important issue in rural areas, and have new planning powers for public transport generally. Their traditional powers to help with infrastructure investment for rural communes are reaffirmed.

Regions are given responsibility for professional training, though this is subject to a specific grant from central government. Apart from this, they are given few service functions of their own, their main role being the planning and economic responsibilities described in Chapter

Seven.

Other functions are devolved to all three levels. In education which remains essentially a national service, communes have the responsibility for building and maintaining elementary schools, the département that for secondary schools and the region that for specialised colleges. All three levels have enhanced responsibilities for planning and location of schools and pupil numbers but decisions on staffing remain in the hands of the state. Mayors have also received new powers over matters such as discretionary holidays and the out-of-hours use of school buildings. In each département there is to be a consultative conseil de l'éducation nationale, bringing together representatives of local government, parents and the state. What this will amount to is that the state has been prepared to hand over responsibility for educational infrastructure and for peripheral matters but the core of the system -curriculum, employment and distribution of teachers, standards - remains firmly centralised.

Ports and navigable waterways have similarly been distributed among all three levels and the state, depending on their size and importance. In cultural affairs too, all three levels will be competent. Housing, as we have noted, remains centralised but all three levels will be more closely associated with decisions in a consultative capacity.

To accompany the decentralisation of functions, the Government has increased its use of the contract system. In opposition, the Socialists had been highly critical of the spread of contracts between the State and local government, regarding them as the instruments of a new tutelle. Now they are being used ostensibly to ensure coherence in a decentralised system but the fears remain on the part of the localities. In culture, for example, where we have seen that decentralisation is to be strictly limited, there are conventions de développment culturel regional through which aid is dispensed to the regions (Queyranne, 1982). In a sphere so cimportant to regionalists, this heralds a continued central control.

It is clear, then, that functional decentralisation is to be strictly limited, with the state closely involved in most fields and the three levels of local government sharing responsibilities widely. Opinions on this differ. Some observers complain that the old idea of 'local affairs' has lost all meaning and that what is being undertaken is not decentralisation but merely a territorialisation of State functions (Burdeau, 1984). Certainly, council's capacity to substitute deliberative and decision-making democracy for the previous 'democracy of access' is not enhanced by the complex interdependence of functions. On the other hand, there are those who claim that any attempt at defining and allocating functions exclusively in the modern state is bound to fail. Belorgey (1984), expressing satisfaction that decentralisation has gone ahead, notes with equal satisaction that, contrary to come of the Socialists' earlier ideas, the unity of the state and the law have been maintained.

Many of the difficulties in arriving at a satisfactory division of functions, of course, stem from the Government's decision to maintain intact all three levels of local government, not to take powers away from any level and not to allow any local authority to exercise a _tutelle_ over another. The resulting complexity and ambiguity gives wide scope for competition among authorities in the same policy spheres. This existed to some degree in the past where, in a city like Lille, the various levels all had their own schemes for improving public transport, no doubt to the great convenience of the Lille travelling public but at considerable public expense. There is also great scope for the _commissaires de la république_ and agents of the central field services to reassert their role in mediating among councils and assembling coherent policy programmes. To that degree, the desire to strengthen political over administrative power may be frustrated. The complex sharing of functions, together with the failure to tackle the _cumul des mandats_, will serve to perpetuate the power of _notables_ able to work on the system at all levels and could serve to frustrate the desire to make government more visible and accountable.

THE MEANS OF DECENTRALISATION

The third dimension to the decentralisation programme involves endowing councils with the means to exercise their new responsibilities. Two items are of central importance here, finance and personnel.

Local government finance in France is as complex a subject as it is elsewhere. While the Socialist Government had committed itself to a complete reform, it was recognised that this would take a long time; meanwhile, there was a promise that transfers of functions would be accompanied by transfers of finance, through a combination of assigned revenues and global subsidies. In 1980, local government was responsible for some 30% of public expenditure, amounting to 9.7% of Gross National Product. Of this, 65% was accounted for by communes and groupings of communes. Communes and départements were raising some 54% of their own revenue through taxes and charges, mainly through four taxes. The most important - accounting for half of tax receipts - was the taxe professionelle. Introduced in 1976, this is based on local payrolls and industrial property values. It is widely criticised as a tax on jobs and for the inequitable distribution of its revenues - according to the level of industrial activity - though over the years equalisation mechanisms have been introduced (Belorgey, 1984). The other taxes, the taxe d'habitation and those on foncier bati and foncier non bati, are property taxes. In addition, councils raised small amounts from various archiac taxes and larger sums from charges. Following a law of 1980 - under the previous government - councils now have the power to alter the rates of individual local taxes and so change the balance among them. Previously, they could only inform the revenue services of the tax product they required and leave it to the latter to adjust the rates proportionally. In practice, the new system has made little difference as it is immensely complicated to change the balance of the taxes, putting it beyond the competence of most communes. Regions had limited taxation powers subject to a ceiling on the amount which could be raised per inhabitant; the amount raised by the regions was in fact very small. Transfers within the local government system were also important. About a third of the communes' income came from subsidies from the départements, which had a responsibility for redistributing resources and helping small communes.

Transfers from central government - accounting for 34% of local expenditure in 1980 - traditionally took the form of specific subsidies for individual investments and services. The conditions attached to these and the need to get approval of plans from the various state services before subsidies could be granted were the basis for many of the financial and technical tutelles under which local councils suffered. Indeed, a whole subsidy culture developed so that councils would put their effort into subsidised investments and avoid others even in cases where the value of the subsidy was negligable. Since the early 1970s, a debate has continued on how to replace these specific subsidies with global grants for investment and services. In 1979, as one of the few practical products of the decentralisation efforts of the Giscard Government, the Dotation Globale de

Fonctionnement (DGF) was introduced for revenue expenditure.
The main components of this are (Belorgey, 1984):
- a guaranteed sum for all councils, diminishing year by year:
- an equalisation element, based on the fiscal potential of individual councils and growing year by year;
- specific elements for categories of council such as small communes or inner city areas;
- a dampening mechanism to attenuate sudden effects of the changes on individual councils from one year to the next.
The amount of the grant, which councils are free to spend in their own way, is linked to the national product of Value Added Tax on the assumption of an unvaried rate and coverage.

Investment has since 1976 benefitted from a global subsidy in the form of the Fonds de Compensation de la TVA. This originated in a demand by councils for a rebate on the Value Added Tax paid on their public works. Given the complexities of this, the State opted for what is in effect a direct subsidy to completed investments, extending it eventually to a range of public bodies such as housing agencies and mixed public-private development companies. The system was widely criticised as open-ended and indiscriminate, subsidising everyone at the same rate and biassing local budgets towards investment rather than service development. The fund has been retained, however, despite the latest reforms.

The major reform of the system of subsidies for investment has been the Dotation Globale d'Equipement (DGE), outlined in principle in the first law, of March 1982, and enacted in a series of laws in 1983. It is being implemented over a period of three years, gradually bringing specific subsidies into a single block. For communes, the principles of distribution of the national total are (Belorgey, 1984);
- at least 70% according to the actual investment of communes;
- at least 15% among small communes according to the length of communal roads, mountainous conditions and fiscal potential;
- the remainder according to the fiscal potential of communes relative to others in the same size bracket, and to districts and urban communities. The original proposals had placed more weight on actual investment. This was changed after objections in Parliament that, by favouring communes which invested most, it insidiously encouraged communal mergers and regroupments.
For départements, the grant is in two parts. The first part

is distributed as follows;
- at most 75% according to actual investments;
- at most 20% according to the length of departmental roads;
- the rest according to the fiscal potential. There is a dampening mechanism to prevent too rapid changes for individual councils.

The second part is distributed as follows;
- 80% according to actual investment in rural areas, including the subsidisation of communes' investments and those of other agencies;
- the remainder according to fiscal potention.

There is no doubt that the introduction of the DGE is a move of great significance, giving councils much greater freedom in determining their investment programmes. Not all the ministries, however, have been playing the game as enthusiastically. New specific funds have been introduced in education, culture and professional training, covering capital as well as revenue expenditure. Like so many of the decentralisation measures, too, it benefits mostly the big councils who are in a position to exploit it. Small communes will need to cooperate with each other and with départements to realise investments. They will also continue to be subject to control by the départements which will allocate much of their investment finance. The evidence so far shows that the départements have been unwilling to globalise their own subsidies to the communes. Some presidents justify this by pointing out that the aid is an instrument of departmental policy, not an automatic entitlement or a means of reducing communal taxation. In some areas, however, it may be the instrument of a new patronage, with councillors able effectively to distribute the subsidies among the communes of their own cantons.

So far, we have been discussing the financing of the traditional functions of local government. For the newly transferred functions, the Government promised new resources, maintaining a balance between assigned or transferred taxes and direct subsidies. The major new subsidy is the Dotation Globale de Décentralisation, a block grant calculated initially on the basis of the cost of transferred services. In future, its growth will, like the DGF, be linked to the product of Value Added Tax. The main taxes to be transferred have been those on the carte grise to the regions and vehicle excise duty to départements, together with some taxes on changes in property use. For the regions, the ceiling on taxation will be lifted when the councils are directly elected. In the meantime, it has been raised, with regions in 1983 taking considerable advantage of this to increase their budgets between 30% and 50%.

While the large percentage rise in regional taxes which this entailed caused some cries of anguish, the absolute levels of regional expenditure and taxation remain tiny in relation to those of the communes and départements.

The final source of finance is loans, which account for some 13% of councils' income and 36% of their investment expenditure. Most of these come from public funds at special rates of interest, notably the Caisse des Depots et Consignations and the Caisse d'Aide à l'équipement des Collectivités Locales. A fairly tight control on these sources of loans and on interest rates is maintained by the Ministry of Finance.

Finance has proved to be one of the most controversial aspects of the whole decentralisation programme, with councils almost unanimously complaining that they have not been given the means to discharge their new responsibilities. The Government, on the other hand, insists that it has scrupulously transferred resources and taxes corresponding to the present cost of services. The problem seems to be that many of the transferred services are ones in which expenditure is growing - for example in social services - while transfers are linked to taxes, including Value Added Tax, whose yield is likely to be restricted in the conditions of austerity anticipated in the post 1983 policy of economic rigeur. It is also alleged that much of the property whose maintenance the state is transferring suffers from a large backlog of neglect. Hence the suspicion that the State is merely transferring the burden of the most difficult and expensive services to the localities. This will only be resolved by a radical reform of the local taxation system and a new approach to the distribution of resources between the State and the localities. It seems unlikely that any government will grasp this nettle for many years to come.

The other main aspect in which localities were to be equipped to meet their new responsibilities was in terms of personnel. Local government officials had long resented their apparent second-class status compared with national civil servants and it was thought that this could affect the prospects of recruiting good staff to carry out the new responsibilities. The solution, embodied in a law of 1984, is extraordinarily centralist. A new corps of territorial administrators is created, grouping most local government officials as well as those officials transferred from central government who choose to join it (most are likely to remain in their existing corps). Recruitment is to be by national examination, with successful candidates allocated to councils according to the preference of the candidate, his position in the competition and the needs of individual council. There is no question of mayors and presidents being able to select individual officials except in the politically sensitive areas of their personal <u>cabinets</u> and for the posts of secretary-general and director of services for councils. Members of the existing national corps of administrators will continue to be able to work in local government and members of the new <u>corps</u> of territorial administrators will be eligible for appointments in central government. Indeed, civil servants transferred to local government will have the option - up to the year 2000! - to return to central government service. Thus, it is claimed, the unity of the national administration will be maintained.

This is a solution very much in the French tradition and clearly extremely advantageous to the officials involved. Local administrators gain their own <u>corps</u> while having free range to step into other parts of government. Control of entry and the allocation of posts remains centralised, with the <u>corps</u> having considerable powers of self-regulation through a new <u>Conseil supérieur de la fonction publique territoriale</u> presided over by a local politician and comprising equal numbers of councillors and representatives of territorial administrators.

It is intended that, as functions are handed over, staff will be transferred from the state services to local councils, though progress on this is proving slow. Central administrators appear anxious less to continue working for the State than to preserve the unity of their <u>corps</u> and keep operational units together, whoever they might be working for. It is likely, therefore that the formula for putting services 'at the disposal' of local government will be more than a temporary expedient. As for the small communes, experience varies greatly. Some fear a new <u>tutelle</u> from using the services of the <u>département</u> and are keeping their links with state services. In other areas, agreement has been reached on setting up departmental agencies to help the communes. In Savoie, for example, such an agency already existed, managed jointly by the <u>département</u> and the communes, who took to it because, unlike the State services in the past, it did not charge according to the size of works involved and try, as a result, to push up their cost. Concern in the Ministry of the Interior about the possible duplication of services, with councils setting up their own agencies parallel to those of the state, has led central government to put some pressure on councils to rationalise these sorts of arrangements.

A PRELIMINARY ASSESSMENT

In November 1983, President Mitterrand declared, in a speech to the prefectoral corps, 'Except to take account of experience, to rectify and adapt where necessary, to specify the means where this has not been done, there will be no more texts on decentralisation' (Regards sur l'Actualité, 102). It will be some years before a mature judgement can be made on the effects of the changes. New roles and relationships will be established on the ground and new 'rules of the game' gradually emerge. As far as the Socialist Government is concerned, however, it would be fair to say that the first dimension, the shifting of power relationships to the benefit of politicians and the local level has been pursued effectively, with the notable exceptions of the postponement until 1986 of direct elections to the regional councils and the failure effectively to establish prefectoral control over the external services of the state; this has hampered the local level dialogue between decentralised and deconcentrated power which was to have prevented the drift of cases up to Paris. Local politicians, however, are relishing their new role, with the political right, previously so critical of decentralisation as a threat to the unity of the state, calling by 1985 for greater powers. Major uncertainties, remain, however, on the political status of each level. The refusal to choose between the levels or to limit the cumul des mandats has preserved some of the most criticised features of the old system, notably the ability of notables to operate at all levels of the system, obscuring accountability. The consolidation of the position of the département, has been extraordinary, considering that this is the level which had earlier appeared most threatened, caught between the region's logic of scale and the commune's deep political roots. A radical reform at its expense is now difficult to conceive for a very long time. Similarly, no attack has been made on the archaic structures of the communes. In the rural areas, these will continue much as before, with some mild incentives to intercommunal cooperation but no effort to establish more viable units - this in itself will further strengthen the départements. Nor is a redrawing of regional boundaries, long demanded by regionalists, envisaged. Vested interests have been created in the existing regions which will only be strengthened with direct elections.

In the cities, mayors will be able to consolidate their

positions but the fragmentation of city government will remain. There has been a minor reform of the organisation and functions of the sever urban communities but these remain essentially groupings of independent communes with no real political life of their own and rather indirect accountability. Weak political leadership in the communities hampers the development of clear policies across the range of urban functions, the mayors jealously guarding their own prerogatives. Where the community is dominated by the mayor of the predominant commune, as has been the case in Bordeaux except between 1977 and 1983, then this is less of a problem. Where there are sharp political differences or the presidency of the community is held by a political opponent of the dominant mayor, severe difficulties can ensue. To resolve these, the community may revert to a policy of <u>saupoudrage</u>, scattering resources evenly among the communes or settle for a consensus style, seeking the lowest common denominator of policy. In either case, the Government's objective of repoliticising local affairs, of encouraging more explicit policy choices and promoting accountability, will be frustrated.

As for the strengthening of democracy within the locality, the programme has done little. Some obserers have predicted the growth of local 'feudalisms' as the big politicians strengthen their grip. Many of these fears are no doubt exaggerated. We have seen that the system remains fragmented; but the strengthening of local associative life and the promotion of <u>autogestion</u> in the community will have to be advanced at the local level.

On the decentralisation of functions and finance, the programme has been much more cautious. Early ideas for a radical devolution of powers have been gradually whittled down as a result of political and bureaucratic opposition. Instead of whole blocks of functions being handed down, ministries have been reluctantly prevailed upon to cede administrative tasks within policy fields and subject to some strict limits of control. Politically, the most important functional transfers may be those in urban planning and development, with mayors now having to take responsibility for giving planning permission. The new -esponsibilities for regional planning and opportunities for economic intervention are discussed more fully in chapter Seven. There we also consider the question of intergovernmental contracts, in the context of the Plan.

Finally, in the matter of providing the means for decentralisation, much remains to be done. It is unlikely that the small rural communes could ever be equipped with the means to take in hand their own affairs; they must rely on support from the départements and the state. For the rest, a reform of local finance is long overdue, though whether any central government or, more specifically, the Ministry of Finance, will be more amenable than their equivalents in other western countries to such a reform is doubtful.

Despite the fears of some earlier critics, the unity of the administrative system has been preserved through the complex interdependence of the various levels, the spread of contracts, the cumul des mandats and the unified bureaucracy. Whether and how such a unified system can cope with the more politicised style of intergovernmental bargaining which is emerging in an issues which we consider in the Conclusion.

6 The change in Corsica

One of the firm proposals of the Socialists in the 1981 election was for a 'Special Statute' for Corsica giving some recognition to a specific (i.e. Corsican as well as French) regional identity, to be formalised by the granting of a directly elected Corsican Assembly and various measures designed to meet indigenous grievances. Moreover, the statute - despite the particulier - was soon to be presented as a blueprint for the rest of France, an initial cog in the wheel of decentralisation and a declaration of good intent. This ambivalence raised certain problems of definition which will be examined below but no case study of contemporary Corsica is complete without some recourse to Corsica's status prior to May 1981. In this chapter, therefore, a short resumé of Corsican history and grievances will provide the background for an examination of Corsica under Mitterrand's regime.

CORSICA AND FRANCE: A 'SPECIAL RELATIONSHIP'

French control of Corsica dates from 1769 following colonisation by different powers, notably Rome, Pisa and Genoa. Genoese rule lasted five centuries and, inevitably, left its mark on Corsican society. Even under French rule, Corsican traditions, social structures, dialect and politcal behaviour had more in common with those of Southern Italy than France. Moreover, French acquisition of Corsica was somewhat controversial. In 1796, unable to maintain hegemony over Corsica, Genoa sold the island to France and thus commenced over two centuries of French Corsica. However, the transition from Genoese to French overlordship was complicated by Corsica's brief flirtation with independence (1754-1769) under the leadership of Pascal Paoli.

The main contemporary significance of the Paolian phase that it provided the twentieth century nationalist movement with a romanticised, idealised, but nonetheless real reference point from which to claim separation or autonomy from France. One notable recruit to the Corsican cause was Jean-Jacques Rousseau, who delighted in Paoli's invitation to him to advise an independent Corsica on a legislative framework. No doubt, Rousseau visualised the embryonic Corsican nation as a potential guinea pig for his social contract. French rule deflated the aspiration of Paoli and Rousseau but, at least, the French Revolution annulled the Versailles Treaty of Annexation, welcomed Paoli's contribution and appointed him commander of the Corsican National Guard - but there was to be no question of separatism, federalism or autonomy. Paoli mistrusted Jacobin zeal and looked primarily to the 'federalist' Girondins. In 1790, significantly, Paoli lobbied not for an independent Corsica but for a <u>statut particulier</u>. Indeed, after centuries of Corsican colonisation, Paoli preferred the protectorate of a major power and Revolutionary France seemed the obvious choice. Paris' reluctance to support this proposed relationship led Paoli to opt for separatism and to solicit British assistance. The latter proved to be short-lived and, before long, Corsica faced 'pacification' at the hands of the island's most famous son, Napoleon 1. Napoleonic France signalled the end of any immediate Corsican prospects for autonomy or separation.

For the contemporary 'nationalist movement', French rule over Corsica represented a continuum with pre-1754 Corsican status. The central platform of the movement rests on the projection of the island as an eternally colonised geographical entity, subject to Rome, the Papacy, Pisa, Genoa enjoying a brief period of independence before returning to colonised status under French hegemony. Arguably, certain features were common in French and to colonial rule - for example, reservation of top jobs for non-Corsicans; development of the littoral and neglect of the interior; suppression of Corsican culture; encouragement of the parasitic clan system; implantation of occupying troops; growth of the Corsican diaspora. Consequently, Corsican experience has lent itself at least to theories of internal or quasi-colonialism and, in part, the French PS has substantiated this broad interpretation by treating Corsica not as 'a region like the others' but as a special case meriting a <u>statut particulier</u>. Regional particularity, of course, is not simply confined to Corsica: for example, as we argued in Chapter Three, areas like Brittany and the French Basque country boast their own specificities. However, neither merit a <u>statut particulier</u> and, notwithstanding the electoral promise to create a Basque dép<u>artement</u>, they have been included in the Socialists' <u>general</u> proposals for decentralisation.

In 1981, the PS accepted that Corsica as a whole had not flourished within the French nation, and Premier Pierre Mauroy even admitted that there were elements of a colonial situation in the Franco-Corsican modus vivendi. The new government's attitude to Corsica reflected the PS's concern to redress the balance and respond to accumulated grievances built up over two centuries of French rule. Critics of Corsica's status vis-a-vis France listed a catalogue of ills, some historic, others comparatively recent, but all apparently attributable to Corsica's constitutional standing. In brief, on the economic level, Corsica experienced premature deindustrialisation, underdevelopment, infrastructural neglect, comparatively higher prices and lower incomes. In agriculture, Corsican small farmers, peasants and winegrowers protested at official policies favouring incoming <u>pieds noirs</u>, the development of the littoral as opposed to the interior and 'big' farming as opposed to indigenous small peasants/farmers. In tourism, the island's main industry, foreign and multinational capital benefitted most whilst the <u>tout tourisme</u> on offer represented a short, sharp interruption of traditional Corsican life-styles. Moreover, the majority of employment vacancies in the tourist industry went to immigrant labour rather than native Corsicans. The influx of migrant labour and <u>pieds noirs</u> was compounded by the swelling Corsican disapora in mainland France and the aging population structure of native Corsicans. Many Corsican emigrants found refuge in the French Army or Foreign Legion, but this military association evoked unpleasant memories: in the first World War, 30% of the Corsican working population died at the front (compared with 7% in metropolitan France). Forthermore, inside Corsica, the presence of Foreign Legion bases, the activities of para-state paramilitaries or secret security forces and the resort of cumbersome judicial apparatus (the State Security Court used to convict Corsican nationalists) conveyed the image of an island under state of siege.

In postwar Corsica, three particular factors are worth underlining since they pave the way to Francois Mitterrand's statut particulier. First, as regards the economy, state intervention under the guise of modernisation served principally to accentuate French hegemony. For instance, the Plan d'Action régionale (PAR) of 1957 proposed the development of tourism and agriculture as the answer to Corsican difficulties. Two government agencies, SETCO (Societe pour l'équipment touristique de la Corse) and SOMIVAC) (Societe pour la mise en valeur agricole de la Corse) were the chosen instruments of modernisation. This attempt to integrate the Corsican economy into mainstream liberal capitalism and international competition mainly benefitted large scale enterprises (for example, Sheraton, ITT, Sofitel, the Rothschild Bank, the Bank of Suez and the Pacquet Company) rather than Corsican hoteliers. Similarly, the main effects of SOMIVAC were to enrich Algerian pied noirs, push Corsican agriculture towards a restrictive monoculture (wine-growing) and, eventually, cause agricultural decline. Unsurprisingly, tourism and agriculture are deemed suitable cases for treatment within the framework of the statut particulier.

Second, the development of Corsican agriculture and tourism was supported by a controversial and important element, the clans. The existence of the clans is beyond any doubt, despite denials to the contrary. A traditional feature of Corsican society, the clans operate like 'a state within a state'. The main clan leaders and fractions are easily identifiable and the clans stand accused of playing a parasitic role between Paris and Corsica. The mediatory role of the clan enabled clan leaders to hunt with the hounds and run with the hares. As dispenser of pensions, favours, jobs and privileges the clan has maintained a hold on Corsican society thereby rendering it more akin to Sicily under the mafia rather than France. In particular, the clans have been attacked by opponents for presiding over the decline of Corsican culture and values. A major source of clan strength has been the duopolism of the system and the concomitant ability of the clans to adapt to the system or regime in power. Hence, Corsica was 'legitimist' under the Restoration, Orleanist under the July Monarchy and Bonapartist under the Empire. Subsequently, via inter marriage or mutual complicity the twentieth century clan had comparatively little difficulty adapting to more recent masters - Mussolini's Italy, de Gaulle,

Pompidou, Giscard d'Estaing and, arguably, even Mitterrand. The electoral changes of 1981, however, and particularly the proposals for Corsica, sowed the seeds of unrest in clan circles. To what extent would the statut particulier and a directly elected Corsican Assembly recast the Corsican socio-political-constitutional edifice? Before turning to this theme, we need to mention another factor - the growth of Corsican nationalism in response to the list of grievances.

The nationalist lobby united diverse movements - separatists, autonomists, regionalists, and reformists. Initially, in the 1920/30s, the movement was based primarily on cultural-cum-intellectual circles and, in due course, suffered from a certain complicity with Italism irredentist claims. More damaging, the movement was tarred with the brush of war-time collaboration. By the 1960s, the movement enjoyed 'a second wind', stimulated by Parisian students and exiled milieux and galvanised by the catalystic side effects of postwar state regionalisation plans. In the Gaullist regime, Corsican nationalism creeped inexorably towards leftist and nationalist perspectives, eventually sprouting separatist bodies (for example, the FLNC, the Corsican National Liberation Front) and autonomist parties (like the Unione di u populu corse). The final pieces in the jigsaw of grievances, state intervention, clanism and nationalism were the Socialist victories of 1981. It is to this area we now turn.

CORSICA UNDER THE SOCIALISTS

The process and institutions of change and decentralisation were not expected to emerge overnight as if by courtesy of some genie's magic lamp. In fact, as we have noted, decentralisation was propagated as 'la grande affaire' of the presidential term of office rather than an immediate turnabout. In Corsican matters, however, the Socialist government demonstrated its intention to proceed quickly with promised reforms.

Under the previous administration of Giscard d'Estaing, false hopes had been raised within nationalist and autonomist circles. Giscard appeared to demonstrate some recognition of the island's specificity and persisted with the traditional policy of throwing money at the Corsican problem. In 1975, the top level mission under Libert Bou went a long way to recognising longstanding indigenous grievances and a charter of redress was compiled. However, Giscard's government failed to support Bou's reformism and, in particular, opposed the concept of a 'Corsican people' or 'personality'. The latter was interpreted as the path to separatism and the break-up of France. In view of Parisian intransigence, the nationalist movement became more radicalised and, in the 1970s, a cycle of demands, violence and repression was evident on the island.

In June 1981, Giscard maintained a right wing victory in Corsica (taking 54% of the vote in Corse-du-Sud, and 50.51% in Haute-Corse) but, overall, the decisive change of government offered prospects for change. Extensive decentralisation and a <u>statut particulier</u> were high on the list of nationalist/autonomist priorities. Thus, the possibility of a <u>Mitterrandiste</u> victory encouraged the FLNC to call a ceasefire and the UPC to claim its part in Mitterrand's success.

The original proposals for a <u>statut particulier</u> derived from the Corsican <u>Parti[socialiste</u> and envisaged the creation of a regional assembly of fifty-nine members, elected by PR and empowered to elect its own executive. The head of the executive would replace the prefect as chief executive for the Corsican region. In addition, there were to be two consultative bodies - an Economic and Social Council and a Council for Cultural Development and the Quality of Life. Powers and resources were to be transferred to the Assembly within the framework of the Plan and National Budget. Further, there were to be regional public agencies or offices to deal with key matters such as agriculture, tourism, employment, land, transport, construction, trade. Finally, there were provisions for Corsican language, radio and television access and academic matters. Basically, much of this format emerged as the PS's project for Corsica after winning office in 1981. Indeed, in 1976-77, Gaston Defferre had unsuccessfully proposed the legislation for a Corsican <u>statut particulier</u> which would grant a measure of Corsican self-rule within the framework of the French nation. Subsequently, the PS became committed to granting powers to a Corsican region. However, the local Socialists within Corsica envisaged greater legislative fiscal and budgetary powers for the island than Paris was prepared to concede. Constitutional impediments and the unfamiliar responsibility of high political office somewhat modified the original proposals but the promise of the <u>statut particulier</u>, a directly elected Corsican Assembly and a recognition of the Corsican 'personality' tempted sections of the autonomist/nationalist movement to participate in the Socialist government's plans to redirect France's relationship with Corsica.

During a visit to Corsica, in August 1981, Bastien Leccia (in charge of 'Corsican affairs') explained the principles underlying the Socialist government's approach to the island: to apply to Corsica the general law of decentralisation; to ensure regional control of the economy via regional, public ownership of the principal means of production and to transfer powers in order to promote self-management and Corsican identity. Of course, the primacy of the French Republic remained intact, but Leccia hoped that this could be reconciled with the Corsican blueprint. Moreover, the Socialists' strategy was enhanced by a number of timely measures favourable to Corsica and designed to elicit nationalist/autonomist participation or support: the abolition of the controversial State Security Court (responsible for convicting many corsican nationalists); the amnesty for political prisoners; the dissolution of the Service d'Action civique (SAC) - the Guallist secret police; the appointment of a Corsican (Bastien Leccia) as Delegate for Corsican Affairs; the promised opening of the University at Corte; and the generally sympathetic and responsive official concern for Corsican concerns.

The Socialist government hoped to defuse separatist tendencies in Corsica by alleviating longstanding grievances. This involved decentralisation, transfer of powers, the recognition of Corsica as a collectivité territoriale - a basic, primary, politico-administrative level of the French state - and the provision of the statut particulier. The latter, passed by the French National Assembly in March 1982, provided for a Corsican Regional Assembly to be directly elected within six months using a proportional representation (by the highest average) system of voting. In addition, there was to be extensive surveillance of the Regional Election to control the customary corruption and personation. Simultaneously, the statut recognised a measure of Corsican specificity and 'personality' by virtue of the island's geographical and cultural identity. These provisions were underlined by the constitutionally elevated legal status for the Corsican region, better provisions for Corsican language and the creation of two specialist consultative bodies to assist the Corsican Assembly - these were the Council for Economic and Social Affairs and the Council for Culture, Education and the Quality of Life. Moreover, the new Assembly was promised budgetary, legislative and consultative roles.

Significantly, the new elected body in Corsica was to be called an Assembly whereas France's other twenty-one regions were to have only a regional council, (conseil régional), whose mode of election remained unspecified as yet. Both the Assembly and the regional councils were proposed as components of the PS's decentralisation plans. A further important attribute of the PS's blueprint for Corsica was the provision for the Assembly to suggest modifications or adaptations to legislation before the French National Assembly affecting the island. This provision enabled the Assembly to engage in dialogue and intercede with the prime minister although it fell short of the aspirations of local Corsican Socialists and others anxious for a fuller, initiating, legislative role for the Assembly. Nevertheless, the Corsicans were also granted a panoply of consultative, specialist, regional, public agencies (or 'offices') to cover such key areas as transport, agriculture, tourism, trade, hydraulics, education and training, culture, regional planning and the environment. As regards budgetary matters, the Assembly ws to enjoy some power to distribute funds within the confines of the national plan and budget. This concession would at least make the Assembly worth lobbying despite the comparatively small sums at its discretion. For instance, its first budget consisted of a block grant of 136 million francs to be divided up by the Assembly politicians. Remarkably, given the disparate political perspectives of the sixty-one members (see below), they were able to agree on the budget without too many problems. However, it is worth underlining that the budget was small compared to money transferred directly by the state for other purposes, for example the 250 million francs to get the Assembly started and 600 million francs for the transport agency ('office'). In any case, a bone of contention and disappointment for the autonomist/nationalist/federalist cause was that final budgetary approval lies with the French government although it is unlikely that Paris will create too many difficulties since it wants the Assembly to work. Thus, in 1983, the government promoted Mitterrand's state visit to the island (see below) as a sign of approval and support for the <u>statut particulier</u> and Corsican Assembly.

The <u>statut particulier</u> divided the nationalist movement and political parties. The FLNC, for example, interpreted this concession as a subtle ruse to soften (but not alter) Corsica's colonised status vis-a-vis France. Consequently, it recommended electoral abstention. In contrast, and despite reservations about the <u>statut</u>, autonomist nationalist forces such as the UPC and PPC (the leftish Corsican People's Party) urged participation. The UPC led the campaign for new electoral registers to eliminate bogus voters and challenge the hegemony of the clans. Anxious to proceed quickly, the government pruned the electoral lists, disenfranchising about five and a half thousand 'voters' - the tip of the iceberg of electoral malpractice according to the UPC! Within the major political parties, there were reservations about the status particulier and the Regional Election. The Gaullist leader, Jacques Chirac, opposed the concept of directly elected regional assemblies and warned against the creation of institutions detrimental to French national unity and Corsican interests. According to Chirac, le probleme corse was not one of institutions but social and economic redress. The Corsican Assembly was therefore doomed to failure since it undermined national unity and did nothing for Corsica. Predictably, Michel Debre shared Gaullist fears about the reak-up of France and treading the separatist or federalist gangplank. Inside Corsica, despite strong reservations, Jean-Paul de Rocca-Serra (the Gaullist clan chief) was careful not to distance himself too far from any initiative which transferred power (and therefore patronage) to Corsicans. The Corsican Gaullists were accused by left-wing opponents of a policy of 'silence and measured acquiescence'. Apprehensive about the statut, Rocca-Serra nevertheless led a combined Gaullist-Giscardian (UDF) - Bonapartist-Conservative (CNIP) list in the Regional Election. The Giscardian UDF (Union pour la[démocratie française), somewhat divided over the statut, largely accepted the umbrella of right-wing unity (after the defeats of 1981), although Jose Rossi led a breakaway UDF list - favourable to the election but against 'Paris'.

United in 1981, the Corsican Left was divided in 1982. The Mouvement des Radicaux de gauche (MRG), minoritaire in contental France but majoritaire in Corsica, presented two lists in the Regional Assembly Election. In Haute-Corse, Francois Giacobbi - the Radicals' clan leader - declined to stand but supported Prosper Alfonsi's campaign 'for a democratic region' - this entailed opposition to an 'imposed' statut, but support for the idea of French regionalisation. On the whole, the MRG remained profoundly attached to republican forms such as the département and was opposed to any proposal which smacked of neo-federalist innovations. In 1981, at a special MRG conference in Haute-Corse, the statut was condemned in favour of a comprehensive regionalisation taking in other French regions (Kyrn, October 1981). Nevertheless, in Corse-du-Sud, député Nicholas Alfonsi headed a second MRG list on a 'Unity and Democracy' platform - more left-wing, unitaire and sympathetic to the Socialists' proposed Assembly. For the Socialist Party (PS), weakly implanted in Corsica, the

Election sowed divisions. Local Socialists protested that
'Paris' had retreated on the initial proposals for Corsican
autonomy within the French nation, leaving the Corsican PS
in a difficult electoral position. Furthermore, the
divided PS was unable to agree upon a leader for the
Electoral list. At the eleventh hour, Ange Pantaloni
(first Secretary of the Corse-du-Sud fédération) carried
official Socialist colours after the withdrawal of Bastien
Leccia and the expulsion of Charles Santoni (ex-first
Secretary of the Haute-Corse fédération, Socialist
parliamentary candidate in June 1981 and an influential
'convert' from the nationalist movement), who led a rival
list. In contrast to Radical, Socialist and nationalist
divisions, the Communists (PCF) paraded characteristic
unity. Dominique Bucchini (Mayor of Sartene and a Member of
the European Parliament) led the PCF's 'Action List for a
New Corsica' - this meant nominal support for the Assembly
and statut particulier, but concern to stress Corsica's
place inside the French nation. The PCF preferred to play
down Corsican particularity within the context of a
constructive programme of industrialisation and reform
(Decider en Corse). The PCF's support for the French
nation was not too unlike the MRG's position although the
local MRG was condemned by the PCF as part and parcel of the
clan system in Corsica.

Besides the main political parties and the various
components of the nationalist movement, a number of smaller
parties, lists or independent contested the Regional
Assembly. These are too many to discuss here: suffice it
to say that the Corsican electorate (201,000) was prsented
with an extensive range of 1037 candidates spread across 17
different lists. With 61 seats available (for six year
duration), many candidates were attracted by the possibility
of gaining the 1.6% of the vote necessary to secure
election.

AFTER THE 1982 ASSEMBLY ELECTION

An immediate significance of the Corsican Election was the
tendency to draw lessons for the rest of France. For
example, Gaston Defferre claimed a victory for the
Socialists' decentralisation proposals. Evidently, the
voters had understood these. To some extent, this was a
reasonable interpretation with an overall turnout (68%)
comparable to general elections. However, the Socialists
were unable to add Corsican Assembly laurels to the national
victories of May-June 1981. The main features of the
Assembly Election were the impressive results for the

autonomists (12.7% of the actual vote, 8 seats); the combined strengths of the right-wing coalition under Rocca-Serra's leadership (28% 19 seats); the disarry of the PS (5.3%, 3 seats); the confirmation of the MRG's leadership of the traditional Corsican political Left (17%, 11 seats); and the continued influence of the PCF (10.8%, 7 seats).

The autonomists celebrated their electoral participation with gains from the Left and the Right, doing particularly well in the northern cities (for example 16.9% in Bastia, 18.47% in Calvi). In theory, the UPC held a potential balance between the Left and the Right in the Assembly, but, initially at least, this was not put into practice. In fact, Prosper Alfonsi (MRG) owed his election as president of the Assembly (and, therefore, head of the executive in Corsica) to the votes of the dissident UDF Giscardians (9.6%, 6 seats). The UPC refused to support Alfonsi. The UPC campaign propagated an 'action plan' rather than an out and out autonomist platform, but a distinctive feature of the campaign was the UPC's denunciation of the asphyxiating duopoloy of the clan system. Consequently, a vote for the MRG was tantamount to a legitimation of the despised clan system. This same policy had been adopted in 1981: prepared to support Mitterrand, the UPC was unwilling to back MRG candidates in the 1981 legislative elections since the Radicals were an integral part of the clans. Understandably, the FLNC was even less accommodating than the UPC. Hence, Alfonsi's election was met by a renewed bombing campaign (on an unprecedented scale) by the separatist FLNC, thereby breaking the truce in force since before Mitterrand's presidential election. At this point, it is worth noting that independence commands little popular support (about 3-4% of the population) for most Corsicans, but FLNC style activity is a force to be reckoned with. As we shall note below, the <u>statut particulier</u> did not terminate separatist activity, although many nationalists were coming round to the idea of some arrangement short of total independence. In the short term, however, 'the change' in Corsica was unsatisfactory for Corsican nationalists.

One not uncommon verdict on the Assembly is that little significant has been achieved in Corsica due to Paris' unwillingness to grant enough autonomy to the Assembly and release sufficient funds to confront Corsican underdevelopment. For instance, the Socialist-orientated Parisian daily <u>Le Matin</u> (13 June 1983) claimed that the Assembly had developed into little more than a super <u>conseil régional</u> whilst the situation in Corsica was more uncertain than ever. Predictably, the UPC (Arritti, 9 June 1983) diagnosed the lesson of the first year as lack of autonomy, which would eventually play into the hands of separatist forces. For Gaullists such as Michel Debre, the danger of separatism was omnipresent in the Socialist blueprint for Corsica. Certainly, the FLNC continued to oppose the Assembly politicians, urge 'decolonisation' and flaunt its policy of 'armed propaganda'. Proscribed in 1982, the FLNC looked to the legal Nationalist Committees (CCN) to voice opposition to the statut particulier, the latest phase of colonialism. The FLNC had demonstrated some willingness to develop its political (as opposed to military) response to French rule. Notably, a ceasefire accompanied President Mitterrand's state visit in June 1983. However, there was no general decline in extraparliamentary militancy and violence on the island (or in continental France). On the contrary, 1982-83 witnessed the emergence of ex-FLNC dissidents and other nationalist paramilitaries, unhappy with the alleged moderation of the FLNC. Racked by internal divisions, the latter lost members and support after the Assembly elections. Violence and selective assassination persisted inside Corsica and, in September 1983, one of the island's leading administrators was victim of political assassination due to his role in 'anti-nationalist offensives'.

The granting of the Assembly was intended to defuse separatist and federalist demands and encourage nationalist autonomist electoral participation. Therefore, the UPC's electoralism was welcomed as a victory for the government's decentralisation proposals. It complemented the honeymoon period following the 1981 Socialist victories. Unfortunately, this phase was short-lived as the FLNC separatist movement renewed its activities and the state turned increasingly towards security matters. The FLNC protested against its 'criminalisation' by the state and the resort to increase repression (more police surveillance, telephone tapping, arbitrary arrests, even 'disappearances') which echoed or surpassed previous right-wing measures. This renewal of the familiar cycle of demands -activity- repression created a difficult environment for the Assembly.

For those who expected immediate, tangible change, government spokespersons were quick to point to years of inertia, underdevelopment and broken promises from predecessors. Bastian Leccia (Corse Le Provençal, 1 March 1983) reminded Corsicans that if change was 'on the cards' they owed it to the Left. This positive attitude was shared by the Corsican PS, which claimed that two years of

Socialist rule had done more for Corsica already than two decades of right-wing politics. Evidently, Mitterrand's victory had 'opened the gates of change for all and particularly for Corsica' (Corse le Provençal, 11 June 1983). The PS's estimation was shared by the PCF, which spoke of 'an historical democratic advance' (Corse le Provençal, 12 June 1983). Even the UPC, despite its impatience and criticisms, continued for a while to support the democratic electoral basis of the Assembly institution. At least, the statut and Assembly realised electoral promises and represented an ouverture, a commitment for change in Franco-Corsican relations. Those expecting rapid change needed to become accustomed to the idea that it would be a 'long march' rather than an overnight revolution (Corse le Provençal, 13 June 1983). In order to underline this reality, Francois Mitterrand visited Corsica in June 1983. President Mitterrand's visit was promoted as a vote of confidence in the Corsican Assembly and statut. Mitterrand accepted that these innovations needed an authoritative boost and his presence in Corsica aimed to lubricate Corsican-Paris relations. Significantly, the state visit was preceded by an economic package from Pierre Mauroy's government intended to facilitate Mitterrand's trip and confirm the Socialists' interest in Corsica. According to an Elysee Palace spokesperson, the President aspired to close the foor on the past and to turn a new page (Le Monde, 14 June 1983). Mitterrand used his visit to address the Assembly, meet local interests and reaffirm his belief in un peuple corse. Of course, Corsica's position within the French nation was established firmly by the President, who promised the financial and technical means to support Corsican development.

According to Mitterrand, "in two years, we have done a lot (for Corsica), but not enough. So let us continue." (Le Monde, 14 June 1983). Of course, as stated above, many observers projected decentralisation in Corsica as a national test. For example, an interested outsider maintained that no "matter how many French people recognise that the island is a special case...the success or otherwise of its autonomous institutions will be seen as a portent if not a precedent for the rest (of France)...What promised to be a great experiment in devolution for France as a whole is threatened, if the Corsican experience is anything to go on, by a lack of generosity and imagination bordering on duplicity. Unless Paris can bring itself to trust the French, much trouble lies ahead." (The Guardian, 16 June 1983).

THE 1984 CORSICAN ELECTION

The above warning proved to be prophetic since, within the next year, the Assembly lunged towards a political stalemate. In 1984, the Assembly became deadlocked when the UPC members adopted an empty chair policy against the lack of powers granted and the right-wing refused to support the second budget. Unable to find a working majority, the MRG-led executive was forced to seek early dissolution by the government, four years prematurely. Before dissolution, however, the government agreed to a 5% minimum electoral barrier for the August 1984 elections. The intention was to reduce the number of electoral lists in the hope of securing a stable majority. To some extent, this strategy worked - only 10 lists (1982:17) and 610 candidates (1982: 1037) contested the elections. Nevertheless, overall, the 5% barrier was still too low to achieve its purpose.

The 1984 elections were marked by the success of two relatively new electoral forces, representing different extremes of the political spectrum. The National Front, capitalising on its good European election performance, achieved 9.22% (6 seats). The Mouvement corse pour l'autodétermination (MCA), successor to the recently proscribed CCN and mouthpiece of Corsican separatism, secured 5.22% (3 seats), marginally ahead of the declining UPC (5.21%, 3 seats). The majority of votes and seats were shared amongst the respective left-wing and right-wing forces with the 'traditional' Left (MRG, PS and PCF) taking 39.70% and 25 seats and the Gaullist-led coalition (29.17%, 19 seats) plus CNIP fellow travellers '7.87%, 5 seats) aggregating 37.04%, 24 seats. The most significant feature of the 1984 poll was the failure of the traditional political forces to command an absolute majority, thereby leaving the king-making (the election of the Assembly president) in the hands of separatists, autonomists and the National Front.

What, then, were the general implications of the Corsican experiment in decentralisation? First, the resort to two inconclusive elections within two years served as a poor curtain raiser for the Socialists' decentralisation programme. Second, the unworkable, ungovernable composition of the Assembly prompted widespread criticisms, especially from the Right, against the use of proportional representation (PR). Did it signify a return to the spirit and practices of the discredited Fourth Republic? Moreover, to what extent would PR and regional elections provide the springboard for extremist representatives, who may paralyse regional assemblies/councils? Third, how 'satisfactory' (prohibitive or, alternatively, fair) was the 5% barrier? Fourth, the provision of regional elections provides another stage for national political debate. Although Corsica, because of its unquestionable specificity, is a bad example to prove the point, regional electoral contests are bound up with national politics. Thus, after the Corsican election, leading opponents of the Mitterrand Government suggested that dissolution of the National Assembly was more imperative than dissolution of the Corsican Assembly. Similarly, the Left's subsequent loss of the Corsican executive was seen by opponents as a national pointer for the Socialist government. Fifth, the initial two years of the Corsican Assembly indicated potential stumbling blocks for the whole decentralisation programme. For example, if Paris is unwilling to decentralise or devolve adequate financial powers to regional bodies, these institutions are unlikely to command respect. On the other hand, if regional chambers are unable to agree on limited budgets due to instable majorities, what is the point of empowering regional executives? In such instances, decentralised powers are destined to go elsewhere by default, for example back to the commissaire de la République (prefect). Of course, it is a vicious circle: regions have to prove themselves capable of wielding powers but, without powers, and/or stable, elected executives how can they provide their worth? Finally, the Corsican experience illustrates that regional layers of administration may function as further appendages to be colonised by the notables in the absence of an effective reform of the cumul. Corsica's two main exponents of multiple office holding are the clan leaders Rocca-Serra and Giaccobi. Unsurprisingly, they dominated the election for the presidency of the new Assembly. Even the possibility of a 'grand coalition' was aired by some political observers. In the event, Rocca-Serra owed his presidency to left-wing divisions and the National Front.

CONCLUSION

Critics of the PS's decentralist blueprint see Corsica as the beginning of the rot for the French nation-state, an unfortunate precedent for other French ethnic minorities. Arch centralists like Michel Debre (a diehard Gaullist and formulator of the Fifth Republic's Constitution) have consistently interpreted regional concessions as the road to federalism and this view is shared optimistically by federalists such as Pierre Fougeyrollas, who adopt the region as the basis of their révolution fédérale, or Robert Lafont (1967) in his call for a révolution régionaliste. For example, Fougeyrollas denotes 'a certain repressive conception of French unity' and makes the distinction between primary and secondary national loyalties. In this context, Corsica is an example of a primary nation within the secondary nation state. In fact, Fougeyrollas sees France more as an empire than a nation, thereby echoing the thesis of 'internal colonialist' critics (such as Lafont). However, the Fougeyrollas, the solution is not the break-up of France but, instead, a more flexible, federal France, capable of accommodating regional diversities in a mutually beneficial relationship. Failure to accomplish a federal revolution will evidently (and Corsican autonomists share this viewpoint) lead to separatism and French decline.

Arguably, too much should not be divined from the Corsican example of decentralisation, which is after all 'a special case'. Of course, the Corsican 'path' remains a victim of its ambiguous status - a specific direction for Corsica and an example for the rest of France. Unlike the rest of France, Corsica at least has a body designated an Assembly and this fulfils a longstanding goal of federalists and autonomists. For Fougeyrollas, the region reprsents the basis of federalism and, therefore, the directly elected Corsican Assembly encourages federalist thinkers to some extent. To date, however, the Assembly lacks a stable political majority, tax-raising facilities, effective legislative intervention, adequate self-control over Corsica and faces persistent opposition from various forces. Further, the clans remain well entrenched inside Corsica (including the new Assembly), thereby evoking the classic French maxim 'plus ça change plus c'est la même chose'.

For Corsica, a major problem is how to combine the declared specificity of the statut particulier with the maintenance of Corsica as an integral part of France and the projection of the Regional Assembly as the first stage of the Socialists' decentralisation proposals which promise regional councils in the whole of France as the next stage. The government appears to accept that Corsica is not une région comme les autres and sympathises with the viewpoint that French-Corsican relations have reflected hitherto a qusi-colonial situation. Rather than perpetuate this unsatisfactory modus vivendi, the Socialists would prefer to re-arrange the relations between Corsica and 'the hexagon'. What the Socialist government was prepared to concede is not so much a divorce but a new marriage contract.

Optimistically, we might conclude that the Corsican <u>statut</u> represents the furthest that the modern French state has been prepared to go along the path of decentralisation. The Corsican proposals rest upon an unprecedented recognition of the island's specificity although separatists continue to protest that Corsican identity remains intact. Consequently, outstanding grievances are severalfold and include: insufficient budgetary and fiscal control; inadequate legislative powers - proposals may be ignored by Paris without contradicting the <u>status particulier</u>; scant recognition of Corsican culture and the idea of 'a Corsican people' - official feet dragging over the promotion of bilingualism; cavalier government management of the Corsican economy - for instance, Mauroy's financial handouts prior to Mitterrand's 1983 visit were welcome but arbitrary, thereby reflecting the traditional manner of Franco-Corsican economic relations; slow transfer of resources and powers. Moreover, the Assembly lacks a stable political majority whilst political observers point to too many layers of government for the island's small population.

Undoubtedly, the Socialist government's proposals involve the ruffling of some feathered nests and further proposals in the pipeline, such as the reform of the <u>cumul des mandats</u>, threaten to assist this process. The arrangements for Corsica, however, rest precariously between the two hurdles - falling short of the federalism, autonomy or independence demands by the nationalist movement, but disturbing traditional Corsican interests and political forces. It still remains to be seen how the dynamics of Franco-Corsican relations will evolve in the 'new' France of the Socialists and whether the Assembly will contribute towards stability, redressing the grievances examined in this chapter and taking the island along the decentralist path.

7 Decentralisation and the economy

ECONOMIC MANAGEMENT AND THE CENTRALIST STATE

Socialism, in France as elsewhere, has long been identified with increased state control of the economy through nationalisation and planning. Socialism's critics, notably on the free-market right, have often accused socialists of being necessarily centralist, by stifling the diversity of entrepreneurial capitalism and the scope for individual liberty which this entails under the control of a monolithic state. The Parti Socialiste did indeed come into office in 1981 pledged to greater state control of the economy notably through a series of nationalisation measures intended to give the state the means to push through the modernisation of French industry and provide the investment needed to meet foreign competition. National planning, which had been allowed to run down under Giscard, would be revived, to provide the guidelines for the major industrial and public investments. It is not surprising, therefore, that many of the Mitterrand government's critics have claimed an essential contradiction between its basic economic strategy and the decentralisation policy. If decentralisation were seen merely as a means of managing social and environmental policy, this would perhaps not matter; but we have seen that the 'girondin' elements in the party saw the decentralisation of economic power as a fundamental element of the whole enterprise. So there is a

109

conflict, at least potentially, between the two strands of policy. If, however, we examine the ideological and practical implications of the two commitments, we can see that there is scope for reconciling them.

In the first place, it is a simplification in the French context to see the left as the promotors of state intervention and the right as the defenders of <u>laisser faire</u> free enterprise. The left does encompass an anti-statist tradition while, on the right, along with a weak liberal, free market tradition, there is a view of the state as the promoter of economic development in collaboration with the private sector. As Hayward (1983, p. 172) puts it, 'French governments had for centuries alternated between policies of passive protection and active promotion - state sponsored capitalism and state capitalism - based upon close collusion between the private sector and its public senior partner.' Under Gaullism, state intervention had been seen as vital to the building of a modern industrial state and the advancement of French prestige. Under Giscard, it is true, there had been a move to a more liberal economic regime in which state protection and subsidies were to be removed and French industry exposed to the winds of international competition. On the left, however, this was not seen as an abandonment of business interests by the state but as a shift of business partners. The Giscardian state was seen as working hand in glove with the big business conglomerates and the multinationals against the interests not only of French workers and farmers but also of small and medium sized businesses. Building a counterweight to this state-big business nexus was a vital task for the left. Trade unions were certainly seen as having a role in this but, given the weakness and divisions of the trade union movement in France, other counterweights were also needed, including in the view of the less jacobin elements, a territorial one.

Ideological expression to this idea was provided in the 'dualist' theory, which we have examined in chapter four. Decentralisation under Giscard was seen as a means of creating a dual state and society by separating state and local matters, retaining economic control at the centre but thrusting social costs onto the periphery which would thus bear the burden of adjustment to the crisis. Equilibrium could only be restored by establishing a political counterweight in the periphery, which could build up its own economic sector (Worms, 1980). This profoundly unmarxist idea was reconciled with the continued rhetoric on the primacy of the class struggle only by the traditional French left wing view of 'capitalism' as referring really only to big capitalism, with small businessmen seen as potential recruits to the progressive alliance, along with workers and small farmers.

We have also (chapter 3) noted the earlier attempt to find an ideological rationale for decentralisation of economic policy and to align the class divide with the territorial one in the 'internal colonialism' thesis put forward by Michel Rocard and the PSU (<u>Parti Socialiste Unifié</u>) in 1965 (Rocard, 1965). This saw the regions of France as colonised by Parisian interests in league, in the last century or so, with capitalism. So the class struggle could proceed in line with the fight for regional emancipation, through people in their localities taking control of the means of production. There was much that was incoherent in the theory, which fitted the <u>gauchiste</u> PSU of 1965 better than the social-democrat Rocard of 1981, but it did at least introduce into left-wing thinking the idea that the class struggle need not necessarily be conducted through the medium of the jacobin state and so made respectable proposals for decentralised economic power.

The Communist Party, too, had been moving slowly and hesitantly away from an exclusive concentration on the centralised state as the instrument of economic policy. In an effort to capitalise on peripheral economic discontents, they emphasised increasingly the uneven nature of capitalist development and the way in which particular regions were exploited. 'Internal colonialism' as an explanation, however, was specifically rejected. That would imply that France was not a nation but a collection of nations exploited by the central one and give credence to the notion that the component parts could one day take their place in the European community of states. The Communist view was much more French nationalist, seeing the opening up of the French economy, notably through the EEC, as responsible for large part of the disadvantages suffered by the regions at the hands of multinational capital (Giard and Scheibling, 1981). A more balanced pattern of territorial development would serve better to integrate the nation and allow all its regions to play their full part. So, while regions and localities should have enhanced powers of economic intervention, this should not be seen as an alternative to state intervention nor as a means for the state to escape its responsibilities.

The question of decentralised economic intervention assumed increasing importance in socialist thinking in the 1970s as a result of a number of more practical considerations. The previous centralised regional policy pursued through DATAR had been effectively run down after 1974 by a failure to increase grants in line with inflation. As mobile investment dried up in the recession and previously booming regions were hit, a diversionary regional policy would have been less and less effective even in the absence of cutbacks in public spending. In any case, dissillusionment had set in with some of the spectacular failures of regional policy like the steel complex at Fos, or with their damaging effects on the local economic and social infrastructure, like the gigantic tourist developments on the Mediterranean.

The changing geography of economic disadvantage hit some areas of socialist support particularly hard. In the 1950s and 1960s, regional policies had been aimed at underdeveloped regions, with little industry, poor agriculture and a high rate of emigration. The economic crisis of the 1970s hit traditional coal, steel and textile areas in the north and east of France, presenting the sort of massive problems of adaptation and restructuring familiar to observers of British regional problems. Table 1 illustrates this clearly, showing regions like Auvergne, Languedoc-Roussillon and Provence-Alpes-Cote d'Azur (indeed the whole of <u>Occitanie</u>) improving their relative economic performance, while Nord Pas de Calais and Lorraine have declined.

Table 1

Gross Domestic Produce per capita relative to national average

France   100

Region	Average 1970-72	Average 1979-80
Ile de France	153	146
Champagne-Ardenne	100	99
Picardie	90	91
Haute-Normandie	108	111
Centre	89	93
Basse-Normandie	77	84
Bourgogne	88	88
Nord Pas-de-Calais	93	86
Lorraine	96	92
Alsace	101	103

Franche-Comte	89	95
Pays de la Loire	83	87
Bretagne	71	79
Poitou-Charentes	72	79
Aquitaine	86	89
Midi-Pyrenees	72	77
Limousin	71	77
Rhone-Alpes	103	98
Auvergne	78	81
Langedoc-Roussillon	69	77
Provence-Alpes Cote d'Azur	88	93
Corse	n.a.	51

Source: G. Sicherman, 'Desequilibres territoriaux', Projet, 185-6 (May-June, 1984).

The response of many left-wing councils to the crisis and the rundown of state aid in the 1970s had been to try and develop their own tools for local economic intervention independent of the state. This, as they discovered, was a legal and political minefield. The basic legal provision was the revolutionary 'Chapelier' law of 1791 which established the freedom of industry and commerce and forbade public intervention. Although this was gradually ignored in its application to central government, the Conseil d'Etat had over the years confirmed its applicability to local government. Only those economic activities having a 'public utility' character and those where private enterprise was found wanting were allowed by the courts though exponents of 'municipal socialism' had often managed to push the limits beyond familiar items such as the supply of electricity, gas and water and public transport to municipal abbatoirs and bakeries. Grenoble which, as have seen, provided something of a model to the Parti Socialiste, had inherited a range of such activities which had resisted both privatisation and nationalisation after the war and the socialist council after 1965 was able to add to them in a revival of the old 'municipal socialism' tradition (Beaunez and Pietri, 1982), running water, gas and electricity services, heating, markets, fairs and abbatoirs, congress halls, transport undertakings and a large direct works department. Generally speaking, however, localities were confined to indirect aid through the provision of land and buildings to industry, though even this was regulated by central decrees.

The crisis of the 1970s brought the issue to the fore with attempts by some municipalities to intervene to bail out firms threatened with closure. In some cases, prefects connived in getting round the law but in some notable cases rescue bids were thwarted. In Besancon, the left-wing municipality had supported the workers at the Lip watch plant in their protest at closure and attempted to zone the land in such a way as to prevent it falling into the hands of speculators. Their déclaration d'utilité publique was refused by the prefect. Later, when they aquired other land for the Lip workers to establish a co-operative, the prefect initially vetoed a 2 million franc grant to buy the redundant Lip machinery (Beaunez and Pietri, 1982). Again in the 1970s, an attempt by the municipality of Marseille to rescue the shipbuilding firm Terrin was vetoed by the prefect. In 1976, the governments' position was stated in a circular from the Minister of the Interior, Michel Poniatowski. This instructed prefects to verify vigorously the public utility character of proposed municipal interventions and the default of private initiative. Three principles were to be observed: the freedom of industry and commerce; the equality of citizens before public charges together with the need to limit public expenditure and protect the interests of tax payers; and the harmonisation of local initiatives with the government's priorities for the aménagement du territoire. Following this, a much more restrictive attitude was taken to local interventions, to the considerable frustration of many local Socialist politicians, who began to give increasing attention to the issue in formulating their own policy programme.

In developing their philosophy of local economic intervention and translating it into practical policy, the Socialists faced considerable obstacles. If local economic intervention was indeed to be a major part of le changement, of the break with capitalism, then local political power would have to be capable of confronting the economic power of a private sector which has itself become increasingly centralised. Further dispersal of political power could risk weakening the public domain, allowing private business interests more scope for playing one area off against another. Increased industrial aid and subsidies would represent a net resource transfer from the public to the private sector and competition among localities to attract industry would benefit disproportionately the wealthy areas able to offer the most attractive incentives. Pressure to intervene for the sake of jobs could be irresistable, though the disposable resources of regions, departments and communes are often tiny in relation to the investment needs of major industries. For example, in the town of Roubaix, it was reckoned that the capital necessary to save the firm of Motte-Bossu, with three hundred jobs, was equal to the entire annual investment budget of the commune (Nouvel Economiste, 331. 5 April 1982).

There was the further problem of possible conflicts between national sectoral policies for industrial modernisation and restructuring, which might imply the rundown of certain sectors, and local needs. From the local point of view, it might be not only politically more tempting but also cheaper, at least in the short run, to keep 'lame duck' (canard boiteux) industries going than to allow them to go to the wall in the name of modernisation and redployment of resources. This whole problem was further complicated by existence of an open economy, requiring French industry to be internationally competitive and by the European Community's rules on industrial aid. For the Parti-Socialiste mainstream, which is strongly pro-EC and opposed to protectionist trade policies, this posed a major difficulty. It was less of a problem for the Communists or the left-wing socialist CERES group who both favour a more autarkic strategy and centralised regional policies.

The new strategy for decentralised economic intervention emerged in the 1970s as the party forged its general decentralisation strategy. The main target of local intervention came to be seen as small and medium-sized firms (PMEs). Apart from the ideological advantage that they were not seen as part of the 'capitalism' that socialism was to transcend, PMEs were more likely to respond to the level of incentives open to localities. So the fashionable 'small is beautiful' philosophy combined with economic judgement to create a policy based more on the growth of indigenous industry in the regions and less on the diversion of major investments. Emphasis was also to be put on preserving existing industrial activity where this had a viable future.

The regions would be the key level for economic development policies. Direct elections would strengthen their political position, their resources would be increased and they would produce regional plans with a strong economic emphasis. Industrial incentives could be decentralised to them from the centre and it was expected that the regions would be able to create powerful development agencies and investment banks to channel resources into local industries and to research and development. Départements and communes would also have greater economic powers, to be exercised in conjunction with the regions, and they would be encouraged to use their existing powers in public investment and infrastructure provision in such a way as to encourage development. The whole system would be regulated, joint strategies formulated and priorities reconciled through a revived, strengthened and decentralised Plan. Regional plans would tie in with the national Plan, drawing together central, local and regional priorities and settling differences through political negotiation among elected politicians.

THE NEW POWERS

After May 1981, the new regime was laid down in a series of laws and circulars, notably the law of March 1982 on 'Rights and Liberties' and the law of June 1982 reforming the Plan. We can conveniently divide the economic powers which regions and local councils now possess into four categories: powers of general encouragement through the provision of land and buildings; direct financial assistance; the power to help firms in difficulty; and the decentralised plan, the means of drawing the various elements together.

Indirect aid to industry in the form of land and buildings has long been the custom. Indeed, in the 1960s and 1970s, many communes saw industrial estates and advance factories as the panacea for problems of development, with the result that there was often global over-provision but a failure to provide the right sort of accommodation for specific types of industry. In rural areas, land was also made available for agriculture, particularly for small farmers. These powers remain but central control of the terms on which land and buildings can be made available has been tightened up, varying from region to region according to the severity of economic problems.

Direct investment aid to industry is available within a nationally-determined framework (La Gazette des Communes, no. 8, 18 April - 1 May 1983). The Prime d'aménagement du territoire is a State grant but is now administered largely by the regional councils. Priority is given to industrial activities creating more than twenty jobs and the grant cannot be combined with the prime regionale à l'emploi (see below). Distribution is based on a national map of three types of zone; (a) the most favoured, with a limit of F50,000 per job created or safeguarded up to 25% of the investment; (b) intermediate zones, with a limit of F35,000 per job up to 17% of the investment; (c) non-aided zones, including the prosperous regions and all agglomerations over 100,000 population. The most important cases and the tertiary sector are dealt with centrally by DATAR.

The prime régionale à l'emploi is available for firms of less than thirty employees and for creations of up to thirty new jobs in any firm. The grant limit is F10,000 in urban areas of more than 100,000 population, F20,000 outside them and 40,000 in areas formerly benefitting from the previous system of special aid for agricultural areas. This is a devolved aid, with the precise criteria and method of distribution determined by the regions themselves.

The prime régionale à la creation d'entreprises is available at up to F150,000 per job for new firms created within the twelve months preceding the demand for aid. The method of distribution is up to the regional councils to determine.

Other forms of direct assistance are available from all three levels essentially in a framework determined by the regions, within rules laid down in the Plan. The major general restrictions are respect for the rules of competition and free enterprise and a prohibition on taking a direct stake in a firm except where it is of the traditional public utility type. Regions can make loans, advances and abatements of interest within conditions laid down by the Ministry of Finance. Départements and communes can participate in these arrangements to complement the region but not on their own initiative. All levels can, however, make loan guarantees to firms within conditions laid down nationally. Communes and départements can give a reduction of the taxe professionnelle in those regions where national regional policy aids are available (so excluding the Paris region).

All levels can make contracts with nationalised industries to help their investment programmes or bring forward work earlier. So early in 1983, the PTT (Post Office) was reported to be in discussion with the localities in Nord Pas-de-Calais about help for its telecommunications modernisation programme.

While the taking of direct equity stakes in private firms is in principle still forbidden, it is possible for regions, departements and communes to set up Sociétés d'Economie Mixte (SEMs) with both private and public capital and for these to take direct stakes. There are already some 600 SEMs in France at the local level and it is likely that this form of intervention will increase in the future.

The other form of direct aid is the power given to all levels to help 'firms in difficulty'. This can be traced to the efforts of socialist councils in the 1970s, notably Defferre's Marseille, to intervene to save jobs in local firms threatened with closure. As the power was really a means of regularising practices that had been pursued, against greater or lesser prefectoral opposition, for some years, it was included in the first decentralisation law of March 1982. Under this clause, all forms of aid are possible except for the taking of direct equity stakes and subject to only two conditions, that the firm should genuinely be in difficulties and that a <u>convention</u> (or planning agreement) is made specifying the mutual commitments of firm and council and the measures to be taken to redress the firm's accounts. A circular (no. 82-102,24 June 1982) from the Minister of the Interior instructed the <u>commissaires de la république</u> on the tests they should apply in deciding whether any given intervention was within the law. These are essentially legal and economic though the room for interpretation of such a vague concept as in difficulty' remains considerable. Having satisfied himself as to the legality of a proposed intervention, the <u>commissaire de la république</u> is enjoined to give every possible help to the council in question, particularly in getting contacts with other relevant agencies such as the Treasury and the Bank of France. More detailed guidance is promised in a manual being prepared for mayors and <u>commissaires de la république</u>. In the meantime, the Government's view appears to be that the power should be used sparingly.

Experience so far shows great variation in the extent to which local councils have chosen to pursue economic development policies. Communes and <u>departements</u> controlled by the right have usually taken the view that the local council should confine itself to the provision of infrastructure and the creation of a favourable climate for industry. Left-wing councils, on the other hand, have in some cases attempted to pursue quite elaborate economic development policies. The <u>département</u> of Herault, for example, under the inspiration of the President of the <u>conseil général,</u> has established an economic development service. This undertakes research into the economic needs of the department and co-ordinates the <u>département's</u> input to the regional plan for Languedoc-Roussillon. It also deals with all requests from the communes for aid towards investment projects, liases with the regional council and central government about their investment projects and handles relations with the series of agencies created by the <u>conseil général</u> or associated with it, such as the land agency, the tourist agency and the <u>Comité d'expansion</u>. In this way, it is intended that the various economic activities of the département can be focussed on coherent policy objectives and related to land-use planning. A new tourist development has been created inland to take pressure off facilities on the Mediterranean coast and a rural improvement project mounted to cover flood prevention, water management, energy generation and agricultural diversification. A policy for helping firms in declining sectors is being mounted and firms in difficulty have been helped. The large public works programme is also being harnessed to the same policy objectives. Despite all this effort, it is realised that the main burden of economic intervention will fall on the regional council, and all the département's specific interventions have been in partnership with regions and communes.

In other cases, where a département does not have a conscious economic strategy, the lead in economic intervention has been taken by other levels. In Haut-Loire, for example, a request from the CFDT trade union for intervention to save a failing firm had to be addressed to central government and to the region, though when a package was put together, the conseil général did agree to participate (Le Monde, 12/5/82).

Among the communes, too, experience has varied. As they already possessed their own executive and were able to create their own services before 1982, there have not been the same administrative reorganisations as in the départements like Herault, though in many cases new economic development staff have been hired. Some interventions have taken place to rescue firms in difficulty but, by and large, mayors have been extremely wary of using this power for fear of setting precedents which could make their task unmanageable. Some communes have stretched the meaning of public utility enterprises to revive the old 'municipal socialism' tradition, while others, such as Bordeaux, take the view that economic intervention is not the task of local government and that such economic activity as is undertaken at local level should be done by the regions.

Regional councils have in all cases taken an interest in economic development and planning, this being one of their principal responsibilities. As in the days before 1982, the scale of their activity has shown great variations. In Aquitaine, the Socialist regional council has set itself the task of retaining employment in agriculture and of encouraging high-technology industries through aid to research and development. It is setting up an 'Institut régional de financement de l'industrie' to monitor developments and examine proposals for aid. In Lorraine, the UDF controlled council has set up the 'Institut Lorrain de participation' to help new firms, new technology and industrial diversification. In line with the council's philosophy, the F10,000,000 capital of this body is provided only 51% by the region. The Champagne-Ardenne, the RPR council is even less favourable to economic intervention and so, in partnership with local councils holds only one third of the capital in the 'Institut régional de participation', the remainder being held by the private sector. In Nord Pas-de-Calais, an economic development service had already been built up before 1982 and this is being strengthened. In nearly all the regions, emphasis is being put on research into local economic problems in order to find the most effective means of intervention.

Considerable difficulties remain in the field of local economic development policies. The new powers of communes and départements, especially to help firms in difficulty, may create pressures and expectations which mayors and presidents are unable to meet. 'Lame duck' lenterprises could swallow large sums of public money with little permanent return and even where a firm has a viable future, the resources of a local council may be well below its investment needs. So it is not surprising that councils, after an early flurry of activity, have proceded very warily; direct intervention by localities is likely to be of only marginal importance and confined to small firms.

Clearly, the main thrust of local economic development policy must come from the regions and we have seen that the new legal and administrative framework is intended to ensure that the major policy initiatives are joint ones by the various levels of government, with the region taking the lead. Much will depend, then, on the quality of the relationships among the levels of government, and where two levels have different view on the desirability of economic intervention at all, problems could be created. The regions themselves must depend a great deal for their effectiveness as agents of intervention on the success of the revival of planning, since it is through setting priorities in their regional plans and influencing the national plan that they can extend their influence. So far, there has indeed been a revival of planning, with national budget allocations following the priorities laid down for the Interim and Ninth Plans. However, in a period of austerity, with increased pressure on public spending, planning becomes a zero-sum game with increased scope for conflict about the allocation of resources. In that case, two dangers present themselves. One is that centralisation will reimpose itself through the plan and the financial system, emptying local discretion of real meaning. The other is that the state and central agencies will seek to impose on regions and local councils the costs of their own investments by inviting them to subsidise schemes. We have noted above the precedent of the PTT in Nord-Pas-de-Calais. In this way, the limited free resources of local councils could be pre-empted leaving nothing for the pursuit of autonomous activities or the development of regional strategies.

The regions also suffer from a continued centralisation of the banking and finance system. Despite the nationalisation of the banks, there has been no attempt to impose decentralisation policies on them or to create the powerful regional investment banks foreshadowed in the Socialists' earlier plans. So there is a limit to the extra resources which the regions can 'lever' to supplement their own restricted funds.

The process of negotiation and implementation of the Ninth Plan would be a test of the ability of the reformed system to cope with the contradictory demands of central economic strategy and regional decentralisation.

THE PLAN

By the time of the Seventh Plan of 1976, dissillusionment with French planning was widespread. Sceptics doubted whether in an economy open to world competition and liable to sudden crises like rises in the price of oil or recession abroad planning was at all possible. The _laisser-faire_ instincts of the Giscard-Barre government served further to downgrade the role of the Plan and the Seventh Plan itself was still-born with the Barre austerity 'plan' also of 1976 making much of it redundant. All that appears to have remained of practical value in it are the interdepartmental Priority Action Programmes to be financed by the state (Green, 1981). The Eighth Plan, for the years 1981-5, had not yet come into operation when the government fell and was replaced by an interim plan of two years while the new government prepared a reform of the system.

As token of the Socialists' commitment to a revival of planning, a separate planning ministry was re-established under Michel Rocard, whose longstanding commitment to decentralisation and regionalism would ensure that these principles would inspire the reformed system. On the other hand, Rocard's isolation in the government as a result of his ambitions and his suspected social-democratic leanings might have reduced the plan's impact in the other ministries. In the event, the separate planning ministry was abolished in the ministerial reshuffle of April 1983, but not before the law on the new planning system had been passed. In 1985, the Planning Ministry was revived to provide a post for Gaston Dufferre after his retirement from the Ministry of the Interior. The circumstances of its revival indicated that it was not to be the powerhouse of economic policy.

While the broad lines of the new decentralised planning system had been laid down in opposition, the detailed elaboration was entrusted to a committee under a Socialist member of parliament, Christian Goux. This delivered its report (Goux, 1982) in June 1982, laying down four basic principles for the Plan :

- to provide the conditions for creative compromises among the various decision-making levels of society;
- to take a long-term perspective;
- to provide for the programming of the totality of public finances and regional schemes.

Regional plans, said Goux, should be neither simple extrapolations of the national plan nor mere bids for extra resources. Rather, they should be purposive development strategies drawing together regional, central and subregional efforts and programmes and organised around three objectives:

- to determine medium and long-term options and objectives for the economic, social and cultural development of the region, taking into account national priorities which impinge on the region;
- to devise programmes of execution by their own activities or through collaboration, notably by contracts, with the state, other local governments, public agencies, private firms and other organisations;
- to pursue a programme of aménagement du territoire, or balanced regional development.

There should be regional 'assises' to consult interested parties and compulsory consultation with départements and communes. On the problem of reconciling regional plans with the national plan, the committee recommended that, while regions should be forbidden to adopt priorities in contradiction with those of the national plan, they should be free to adopt different ones, though foregoing aids tied to the national priorities. Implementation of the plan would be through contracts and financial inducements.

The planning process would be in three phases. Firstly, there would be a consultation of the regions to ascertain their strengths, their weaknesses, their priorities and their needs. Then, furnished with this information, a new national planning commission, bringing together the 'social partners' and the state, would make national priorities known to the regions. Finally, each level would draw up its plan in partnership with other levels of government and non-governmental interests.

The Plan itself would be voted in two stages. A first loi d'orientation would lay down broad national priorities and seek to harmonise and reconcile regional plans. A second law would then deal with the implementation of the Plan, obliging the state to provide the necessary resources in its annual finance bill. Any failure to do this would have to be the subject of an amending law. This, together with the extensive use of contracts, would give the Plan more effective 'teeth' than it has had in recent years.

The Goux report was immediately adopted by the government and its proposals formed, in essence, the new law under which the Ninth Plan was prepared. This involves an elaborate series of interchanges between national and regional levels, giving rise eventually to a series of planning contracts between the State and the regional councils. In the first phase, in January 1982, the regions formulated their priorities and informed the government which itself transmitted its broad priorities to the national planning commission. In December 1982, the commission presented its report. following which the government prepared its first panning law and the regions their plans. While the first planning law was being passed, the regional plans were examined in the CIAT (Comité interministériel pour l'aménagement du territoire) which proposed any necessary amendments and sent the plans back. After further advice from the national planning commission, the second plan law was passed in late 1983 while at the same time the regions were adopting their plans. The final stage was the negotiation and adoption of planning contracts between regions and the state (Durand, 1984).

The main emphasis of the ninth plan is on modernisation and the needs of industry. Financial commitments are made through twelve Programmes Prioritaires d'Execution (PPEs) whose objectives were expressed in very general terms as follows:

1. Modernise industry with new technologies;
2. Renew education and youth training;
3. Encourage research and innovation;
4. Develop the communication industries;
5. Reduce energy dependence;
6. Encourage employment;
7. Sell France abroad;
8. Produce a good environment for the family;
9. Make a success of decentralisation;
10. Improve the quality of urban life;
11. Modernise and improve the management of health services;
12. Improve the system of justice and security.

Sub-programmes spelt out in more detail the financial commitment to specific measures, the bulk of finance going to PPEs 1,2 and 3(Le Garrec, 1983). In the 1984 budget, PPE measures were to receive a 16% increase, compared with 6.2% for state spending as a whole. Even when an emergency spending cuts package came in March, ministries were instructed to exempt PPE items, finding the savings elsewhere in their budgets (<u>Regards sur l'Actualité</u>, 103).

In setting out their own priorities, the regions have generally shown a willingness to accept their new role, though the opposition-controlled Ile de France roundly told the government that the management of the economy is and should be a central responsibility and that, even if were it appropriate for them to attempt it, the regions lacked the means to take in hand their own economic development (DATAR, 1982). Midi-Pyrenees, on the left, predicted that the ninth plan would be as empty as its predecessors unless there was a profound change in the relationships among the state, the localities and industry and a radical reshaping of the means of intervention. The depth of analysis which the regionss brought to their plans varied considerably. Those with well-developed planning and economic services were able to present quite sophisticated analyses of their needs, resources and prospects. Others were more dependent on the services of the state for data and presented little more than a shopping list. Eighteen of the twenty two regions stressed the needs of agriculture as a priority; as this was not a priority of the national plan, it presented a test of the new system's ability to reconcile national and local needs. On the other hand, most regions also stressed the needs of modernisation and, particularly, their desire for a share of high technology and communications developments.

In the negotiation of the planning contracts, the distinction was kept clear between programmes corresponding to PPEs and programmes 'compatible' with the national plan. The bulk of state financial support, it was made clear, would go to the former. The latter could, of course, be financed by the regions from their own resources, but these were likely largely to be preempted in the 'cofinancing' of PPE measures in an effort to get the maximum money out of the State. So planning contracts could easily turn into a mechanism of recentralisation, a means of tapping the regions own resources for the implementation of national programmes. This is all the more likely in that the regional planning contract is in reality a series of contracts with individual ministries who are anxious to keep funds to their own functional programmes. For the purposes of execution, indeed, this reality is recognised as each regional contract gives birth to sub-contracts in specific fields.

In the event, this proved to be something of a problem but not a serious obstacle to the negotiation of the contracts. Some money was available from interministerial funds such as CIAT (fonds interministériel pour l'aménagement du territoire) for 'compatible' programmes, so that just 56% of State expenditure in the contracts in 1984 will be on PPE items. For the regions, commitments to cofinancing through contracts will account for about 40% of their expenditure in the same year. For the five-year duration of the plan, the State's contribution to contract expenditure is to be F35 billion and the regions' F27 billion (Regards sur l'Actualité, 103). This, however, does not allow for any further contracts, in the field of culture or elsewhere, which might be negotiated in the future.

Another major difficulty - apart from the recentralising potential of the contract system - has been the lack of leverage by the regions over the operations of nationalised industries. As nationalisation has been such a prominent part of the Socialists' economic and industrial programme, the absence of a link with decentralisation is remarkable. Many of the regions have expressed misgivings about this but, while the government has conceded that it is perfectly legitimate for the regions to engage in a dialogue with the state enterprises on their territory and legally possible for them to contract with them (DATAR, 1982), in practice it has been made clear that the orientation of nationalised industries is a national matter. The national priority is to make them efficient and competitive and regional initiatives are not be allowed to prejudice this.

Another potential problem concerned the relationship between the negotiation of planning contracts and the national priorities for regional development. It is possible through the system for the larger subsidies to go to those regions most willing and able to cooperate in the achievement of national objectives rather than those whose development - on grounds of need - is the greatest priority. In fact, PPE is largely about overcoming this. 83% of the expenditure under this heading is for 'corrective' measures of regional policy, mostly to be administered by the state through DATAR, though some of it goes through the regions. The same programme also specifies the national hierarchy of priorities for aménagement du territoire. The remaining moneys under PPE 9 (for 'making a success of decentralisation') are for the re-equipment of prefectures and administrative tribunals with staff and communication technology and for rural transport. A better distribution of State aid to the localities through the Dotation Globale de Fonctionnement and other instruments (see chapter 5) is promised in the same section, though this is not part of the finance of the plan itself.

While Nord Pas de Calais shows the potential for planning in a homogeneous, well-intergrated region, Languedoc-Roussillon illustrates the problems faced by the more articifial creations. Languedoc differs economically as well as culturally from Roussillon and the regional council lacks a model of the local eocnomy or adequate data. So despite an extensive process of consultation, most of the options for consideration in the regional plan were in practice formulated by the central state services, to correspond to the national priorities. The result is a series of sectoral programmes, with credits distributed among the départements according to the old principle of saupoudrage. This is not to say that there has not been a strategic choice, albeit implicit rather than explicit. The strategy is a continuation of that of previous governments, for modernisation of the local economy and its insertion into the open French and European system, albeit with some help for declining sectors. Having vacillated in the 1970s between defence of viticulture and support for transformation, the Socialists, in power at central and regional levels, have opted for the latter, with the network of notables acting as a shock-absorber. Between the integrationist thrust from national and European levels and the traditional local network of notables, there is little room for a regional vision of autonomous development.

A specifically Occitan content could be given to the modernisation strategy with a cultural and linguistic policy but the same process has been apparent here. Cultural decentralisation has tended to mean the decentralisation of French culture rather than the promotion of regional cultures. Of the seven cultural priorities in the planning contract with the state, the only one with no money attached is that for the promotion of the traditional cultures of the region. This, of course, outrages the Occitanists but the council's view is that Occitan language and culture are a minority taste, especially in the cities; to which the Occitanists reply that this is because they are not promoted.

The experience of regional government so far has thus been a great disappointment to the Occitan movement, but it is symptomatic of that movement that the response has been disjointed and often incoherent. With the retreat of the occitanists from the political front line, the movement has tended to fragment into its intellectual, cultural, economic and political components. An organisation called SPELEO (Société pour l'étude de l'espace occitan) produced a critique of the plan, on the basis of a conference (SPELEO, 1985). This revealed most of the strands of criticism of the Socialist version of regionalism from the Occitanist point of view, but also the failure to agree on a coherent alternative. There was an argument between the Communist-led CGT trade union and others over protectionism, with the CGT representative advocating a large measure of regional autarky. Paradoxically, the CGT and Communist Party's espousal of this line appears to bring them close to the fundamentalist nationalists of VVAP, accused of a 'virage a droite'. In fact, it reflects a combination of the Communists' support for protectionism at the French level with the local Communist Party's resolute defence of existing producers. This does not make the Communists Occitan nationalists for they are in favour of regional government only in the context of a strengthened French state, outside the European Community (Giard and Schiebling, 1981). Most 'progressive' Occitanists deplore this attitude and seek an outward-looking, modernising regions; yet this, as everyone knows, will mean the end of the traditional viticulture and the way of life which goes with it.

There have been widespread complaints that the unions had not been sufficiently consulted on the plan. There are complaints about the neglect of regional culture; but nowhere is there a coherent counter-plan, reflecting a realistic alternative view on the region's future.

The Socialists have done most of what they promised to do in terms of increasing the economic powers of local government; but, like other aspects of their programme, the content of the policy proposals was less radical than the rhetoric in which they were clothed. It is difficult to see local economic interventions as they have developed as in any sense a 'rupture' with capitalism. Interventions follow the logic of the market economy and are intended to sustain it. Nor is local intervention likely to result in a substantial decentralisation of economic policy making itself. Certainly, some of the regional policy aids formerly administered by DATAR and the prefects have been decentralised to the regional councils, but these aids are awarded more or less automatically to private firms, within a national framework of rules and could certainly not be used to forge a radically different approach to economic activity. More opportunity for creative activity may be available through research into local economic problems, especially into economic linkages and sectoral gaps. Resources, in the form of direct aid and infrastructure provision can then be concentrated where they are most needed and on firms and sectors which will respond to the level and type of incentive provided. To a large extent, though not entirely, this will involve concentrating on small and medium sized firms, leaving the major cases to central government. So we are not witnessing a radical decentralisation of economic intervention but the putting into place of a new level of intervention, parallel to the centre's policy but not in contradiction to it. The recession has given ample scope for such increased intervention.

As with other aspects of the decentralisation programme, the true extent and significance of local economic intervention will not be clear for some years; indeed its real test may not come until a right-wing central government faces socialist councils trying to pursue interventionist strategies against the grain of state policies.

8 Conclusion

We have described the administrative effects of the Mitterrand government's decentralisation programme, showing that real measures of decentralisation have been undertaken and that there has been a major shift in power relationships, albeit one which reinforces existing trends. The assumption of executive powers by the President of the conseil général marks a radical change in many départements and the greater political exposure of local politicians following the removal of the prefectoral 'umbrella' could have significant effects on behaviour. Too cynical an interpretation of the programme, on the lines of plus ça change, plus c'est la même chose, would not, therefore, be in order. On the other hand, vital features of the traditional order have been maintained. We have noted the consolidation of the département and the survival of the small communes. With the exception of the Corsican reforms, anything which could remotely be suspected of encouraging structural change in the system has been avoided. Consequently, many of the hopes of regionalists have been disappointed. The cumul des mandats, cornerstone of the notable power structure, has been left intact while the complexity of the division of responsibilities and the grudging manner in which ministries have been persuaded to hand down functions, piecemeal, militates against clear local responsibility for policy and its implementation. Despite the Socialists' early rhetoric about a new economic order, functional decentralisation has largely been confined to social and, to a more limited extent, environmental policy, with both the party's residual jacobinism and its espousal of economic orthodoxy after 1982 preventing bold moves on the economic front.

So the system remains interdependent, a local democracy of access rather than one of deliberation and decision. To some degree, this may be inevitable in the modern state. Local affairs are no longer easy to distinguish, conceptually and practically, from national affairs, given the imperatives of economic management and of the welfare state with its commitment to equity and national solidarity. We do consider, however, that more could be done to clarify the role of each level of government and its relationship with the centre. This would mean simplifying the system and seizing the nettle of structural change. The region is widely regarded as the most suitable unit for planning and the provision of many decentralised services, though in some cases boundaries are in need of changing. The commune, despite the existence of many tiny communes whose functional viability is very much in doubt, retains a vitality as the expression of a localist democracy which has disappeared in mnany other parts of Europe. The département can be justified on neither ground. A technocratic creation of Napoleon, it neither serves the needs of modern administration nor provides a basis for real participation. A truly radical reform of French territorial government would involve an attack on it and on the political class which sustains it. Only with an assault on the power of the notables could the vision of a 'new citizenship' begin to assume reality.

Structural change is, of course, only one element determining the future balance of territorial power. Political and behavioural factors will be of equal importance in moulding future patterns. Since the brave days of 1981 there has been a profound change in ideoligical fashion in France. Liberalism, taken up enthusiastically on the right, has been enjoying a vogue in intellectual circles and even affected the *Parti Socialiste* and the Mitterrand Government. Since 1983-4, the message from government has been on the need to heed the market, to encourage competition and to rely less on the state. At first sight, the new liberalism might appear conducive to decentralisation and, to some extent, it draws on the same reaction to the stifling influence of the centralised state. On the other hand, the liberal vision focusses on the freedom of the individual and the firm, not that of the collectivity. Indeed, in its anti-collectivism, it may be the enemy of the decentralised visions of 1981 and the autogestionnaire ideals inherited from 1968. These centred on the local community and its capacity collectively to manage its own affairs.

At the same time, there has been something of a re-emergence of jacobinism. Jean-Pierre Chevenement, leader of the left-wing CERES faction had the opportunity to push his jacobin ideals after replacing Alain Savary as Minister of Education in 1984. To the horror of observers such as Edmond Maire general secretary of the CFDT and a leading proponent of autogestion and decentralisation, progressive education began to be abandoned, central contol reimposed and discipline and competition encouraged in the name of 'republican elitism' (Nouvel Observateur, 4/1/85). The ideal appears to be a return to the traditions of Jules Ferry and the state education of the Third Republic, with its emphasis on patriotism, work and reward. The small measures of decentralisation undertaken under Savary to emphasise the needs of 'society rather than the State' were reversed and central control over curriculum and school organisation tightened.

Between the market discipline of the new liberalism and the reassertion of the jacobin state, the prospects for a new type of decentralised political, social and economic order may not look promising. We have seen in Chapter 7 that there has been little decentralisation of economic policy. Rather, a parallel set of interventionist measures has been put in place at the local level. The joint pressures of renewed jacobinism and market liberalism will make the achievement of local economic strategies all the more difficult. As part of this programme of rigueur and economic 'realism', the Socialist Government imposed tough commercial targets on the nationalised industries. There is no question of their responding to local pressuress to preserve employment or restructure their operations in accordance with the needs of local economies. Local control of them is thus as far off as ever. Nor have the major economic interest groups turned much attention to the potential of local government. Certainly, the CFDT trade union remains a firm supporter of decentralisation and regionalism but neither the other unions nor the employers' organisation, the CNPF, have shown much interest. Rather, they continue to look to the state for help with their difficulties. Much of this, of course, was foreseeable. The implicit contradiction between nationalisation and state planning and decentralisation was never convincingly resolved in the Socialists' early plans; rather, it was merely denied with the assertion that a more political and contractual style of planning would reconcile state and local priorities. The recession which has replaced the expected economic growth has compounded the difficulties. At the same time, however, it has confirmed the need for local responses to the problem of unemployment and the pressure on local government to provide these.

Faced with the alternatives of an economy totally open to international competition, with its potentially devastating effects on disadvantaged regions and localities, and the 'seige economy' favoured by sections of the left in France as in Britain, the French socialists opted for a middle course. This proved immensely difficult in practice and the problems are likely to be accentuated with the renewed emphasis on international competitiveness as the major aim of policy. The logic of the pre-1981 Socialist position, indeed, was a type of 'dualism' of the sort which they themselves so roundly condemned, with an internationally competitive industrial sector and a sector of industries and public and private services whose justification would be found in their local and regional cost-benefit ratio but which would have to be protected from international and, to some extent, national competition. Such sectors would tend to be locally and regionally-based to avoid encouraging the <u>chasseurs des primes</u> and the subsidisation of industry to relocate.

There is some evidence that some of these elements are being put in place in the more dynamic and progressive regions in spite of increasing central restrictions. What is lacking, however, is locally and regionally-controlled investment finance. The nationalisation of the banks has not changed their attitudes towards the PMEs or local businesses. There is also a question mark over the direction of regional plans. Though the existence of the regional councils has certainly led to a growth of regional awareness, the delay in direct elections gives cause for concern. As long as regional councils are dominated by representatives on the <u>départements</u> and communes, there will be a strong temptation to a policy of <u>saupoudrage</u> rather than the forging of coherent regional strategies. The election of regions on a departmental list system in 1986 will paradoxically mean that even direct elections could serve as a reinforcement of the <u>département</u> as the basic unit of the local powr system. Here again, experience differs. Some regions have long concentrated their investment subsidies to gain the maximum amount of leverage. Others, however, even while rhetorically proclaiming the goals of modernisation and change, are scattering their resources across the traditional types of activity.

On the broader political front, the prospects for decentralisation may be better. The right wing opposition, which began by attacking the whole programme, from about 1984 began pushing for it to be speeded up. The very fact that the programme consolidates the power of existing elites guarantees it wide support from the political class and opposition mayors and presidents of regional and departmental councils lost little time in exploiting their new powers to the full. If they lose the legislative and presidential elections in 1986 and 1988, the Socialists in turn are likely to fall back on their territorial power bases. For all the talk of Jacques Chirac, the right-wing government would be unlikely to repeal the main legislative provisions of the decentralist programme. If this does represent a recognition in France of the legitimacy of the alternation of power and of its territorial distribution, that will mark a considerable change. It is true that, following the Socialists' local reverses in 1982 and 1983, the Opposition tried to claim that the Government had lost its democratic mandate. This was the classical jacobin pose. The republic being indivisible, if the Government lost elections at any level, that must represent a withdrawal of the people's confidence. By 1984, they had calmed down somewhat and appeared content to wait until the proper time for legislative elections, using their local power in the meantime. Proportional representation for national elections will represent a further move to pluralism and, possibly, a weakening of the centre, and the system of lists by département will reinforce that institution yet further. In a Parliament without a guaranteed government majority, local pressures could be greater and the scope for local independence greater.

In the longer perspective, recentralising pressures are likely to come from national politicians, from the bureaucracy and from the fiscal crisis of the state. Politicians of all persuasions tend to cling to their national programmes, even where these do not impinge on economic management, as we have seen in the case of culture and education. It will take an assertive local government system and a sympathetic Ministry of the Interior and Decentralisation to combat this tendency; although Pierre Joxe, Defferre's successor in the 1984 reshuffle, proclaimed his attachment to the decentralisation programme, he has generally been regarded as a traditional left-wing jacobin rather than an adherent of the new girondin wing of the party. The traditional <u>notable</u> is quite content to work through the mechanisms of the centralised state, using it to gain special treatment for his own area. If decentralisation is to work, politicians must emerge demanding the right to take their own decisions and accept the political consequences. There is some evidence that such politicians are emerging - indeed this was one of the factors which brought decentralisation onto the agenda - but they need the encouragement which would be given through direct elections to the regional councils, with a ban on accumulating regional presidencies with national office. Only thus can a viable regional political system be created, centred on the institution of the regional council and not on its links with central ministries.

As we saw in chapter 6, the Corsican experiment offered little joy as regards the <u>cumul des mandats</u>. On the contrary, the 1984 Corsican Assembly elections were seen widely as the 'revenge of the clans', with local <u>cumulards</u> re-establishing and strengthening their position, temporarily disturbed by the <u>statut particulier</u>.

The bureaucracy are a further centralising force. We have noted the way in which they have used the opportunity of decentralisation to create a unified national <u>corps</u> of territorial administrators with considerable privileges. They, often with the support of ministers, have also dragged their feet on the deconcentration of power to the <u>commissaires de la république</u>. Dossiers continue to find their way to Paris for decision, weakening horizontal co-ordination and control at the local level. It will require a continuing exercise of political will to reassert political control and, especially, local political control.

Fiscal pressure is likely to be one of the greatest recentralising forces. In the absence of a thoroughoing reform of local finance, councils will find most of their resources pre-empted for the discharge of their statutory duties, leaving little room for choice or innovation. Continued retrenchment in public expenditure, which is the prospect under with the Socialists or the right, could fall heavily on the localities, with the withdrawal of central support while the costs of providing services increase. Had decentralisation occurred in the days of expansion, the oppportunities for diversity and innovation would have been legion. As it is, there is a risk that local government will continue to be merely an administrative arm of the central state.

Decentralisation came onto the political agenda in France partly for ideological reasons but partly because of the recognition that centralisation was sapping the capacity of the French state itself for adaptation and change. Decentralisation could not only free the localities from the centre; it could free the centre from the pressures of local politics. Thoughtful politicians, tempted to recentralise their practices if not the legislative framework, may remember this and refrain. Political habits, however, take a long time to change and decentralisation is only in its infancy. Already, some options have been closed with the refusal to reform basic structures. Others are still open, given the political will to make the system work and the emergence of a more pluralistic culture. If that can be achieved, then decentralisation may indeed be recognised by history as the grande affaire of the Mitterrand Presidency.

Bibliography

Ardagh, J. (1982), <u>France in the 1980s</u>, Penguin, Harmondsworth.

Ashford, D. (1982), <u>British Dogmatism and French Pragmatism</u>, Allen and Unwin, London.

Barelli, Y, Boudy, J-F and Carenco, J-F, (1980), <u>L'Espérance Occitane</u>, Entente, Paris.

Beer, W. R. (1977), 'Social Clas of Ethnic Activists in Contemporary France', in Esman, M. (ed.), <u>Ethnic Conflict in the Western World</u>, Cornell University Press, Ithaca.

Bell, D.S. and Shaw, E, (1983), <u>The Left in France</u>. Spokesman, Nottingham.

Berger, S. (1977), 'Bretons and Jacobins: Reflections on French Regional Ethnicity', in Esman, M. (ed.), <u>Ethnic Conflict in the Western World</u>, Cornell University press, Ithaca.

Bernard, P. (1983), <u>L'Etat et la décentralisation</u>, Documentation Francaise, Paris.

Birnbaum, P. (1977), <u>Les Sommets de l'Etat</u>, Seuil, Paris.

Boujol, M. (1969), Les Institutions régionales de 1789 à nos jours, Berger-Levrault, Paris.

Crozier, M. (1964), The Bureaucratic Phenomenon, Tavistock, London.

Charzat, M. (1981), 'Etat, Service Public, Autogestion', in La Democratie en Jeu, Club Socialiste du Livre, Paris.

Chassagne, Y. (1983), 'La Décentralisation : Des Rapports Nouveaux entre l'Etat et les Collectivités Locales', Nouvelle Revue Socialiste. 62.

Chevallier, J. (1982), 'La Réforme Régionale', in Chevallier, J., Rangeon, F. and Sellier, M. Le Pouvoir Régional, Presses Universitaires de France, Paris.

Cobban, A. (1965), A History of Modern France. Volume 3: 1871-1962, Penguin, Harmondsworth.

Crozier, M. and Friedberg, E. (1979), L'Acteur et le Système, Seuil, Paris.

Dearlove, J. (1979), The Reorganisation of British Local Government, Cambridge University Press, Cambridge.

Dubet, F. (1983), 'Des Nations sans Etat au pays de L'Etat', Autrement, 47.

Duhamel, A. (1982), La République de M.Mitterrand, Grasset, Paris.

Dulong, R. (1975), La Questionne Breton, Presses de la Fondation Nationale des Sciences Politiques, Paris.

Dupay, F and Thoenig, J-C (1983), 'La loi du 2 mars 1982 sur la décentralisation', Revue francaise de Science Politique, 6.

Dupuy, F. and Thoenig, J-C (1985), L'Administration en Miettes, Fayard, Paris.

Frears, J.R. (1977), Political Parties and Elections in the French Fifth Republic, C. Hurst, London.

Feurer, L. S. (ed.), (1959), Marx and Engels. Basic Writings on the Politics and Philosophy, Fontana, Glasgow.

Giard, J. and Schiebling, J. (1981), L'Enjeu Régional, Editions Sociales, Paris.

Gontcharoff, G and Milano, S. (1983), La Décentralisation. Nouveaux Pouvoirs, nouveaux enjeux, Syros, Paris.

Gourevitch, P. (1977), 'The Reform of Local Government in France : A Political Analysis', Comparative Politics, 11.

Gourevitch, P. (1980), Paris and the Provinces, Allen and Unwin, London.

Gravier, J-F, (1972), Paris et la Desert Français, 2nd ed., Flammarion, Paris.

Grémion, P. (1976), Le Pouvoir Periphique. Bureaucrates et notables dans le système politique français, Seuil, Paris.

Grémoin, P, (1981), Régionalisation, régionalisme, municipalisation sous la Ve republique', Le Débat, 16.

Grémion, P. and Worms, J-P, (1975), 'The French Regional Planning Experiments', in Hayward, J. and Watson, M, (eds.), Planning, Politics and Public Policy: The British, French and Italian Experience, Cambridge University press, Cambridge.

Guichard, O, (1976), Vivre Ensemble, Documentation Française, Paris.

Guillorel, H. (1981), 'Le Mouvement Breton', Pouvoirs, 19.

Hayward, J. (1969a), 'From Functional Regionalism to Functional Representation in France : The Battle of Brittany', Political Studies, XVII.

Hayward, J. (1969b), 'Presidential Suicide by Plebiscite : de Gaulle's Exit, April 1969', Parliamentary Affairs, XXII.4.

Hayward, J. (1983), Governing France : The One and Indivisible Republic, 2nd edition, Weidenfeld and Nicolson, London.

Hechter, M. (1976), Internal Colonialism. The Celtic Fringe in British National Development, Routledge and Kegan Paul, London.

Johnson, R. W. (1981), The Long March of the French Left, Macmillan, London.

Keating, M. and Bleiman, D. (1979), Labour and Scottish Nationalism, Macmillan, London.

Lafont, R. (1967), La Révolution Régionaliste, Gallimard, Paris.

Langumier, J-F, (1983), 'L'aménagement du territoire a-t-il un futur ?', Autrement, 47.

Lebèsque, M. (1970), Comment Peut-On Etre Breton? Seuil, Paris.

Lijphart, A. (1977), 'Political Theories and the Explanation of Ethnic Conflict in the Western World : Falsified Predictions and Plausible Postdictions', in Esman, M. (ed.) Ethnic Conflict in the Western World, Cornell University Press, Ithaca.

Machin, H. (1977), The Prefect in French Administration, Croom Helm, London.

Machin, H. (1979), 'Traditional Patterns of French Local Government', in Lagroye, J and Wright, V. (eds.), Local Government in Britain and France, Allen and Unwin, London.

Mény, Y. (1974), Centralisation et Décentralisation dans le Débat Politique Français, 1945-69, R. Pichon et R. Durand-Auzias, Paris.

Miliband, R. (1977), Marxism and Politics, Oxford University Press, Oxford.

Paillard, B. (1983), 'Fos: "La Villette" du Midi?', Autrement, 47.

Parti Socialiste (1980), Project Socialiste, Club Socialiste du Livre, Paris.

Parti Socialiste (1981), La France au Pluriel, Entente, Paris.

Phlipponeau, M, (1980), Décentralisation. La Grande Affaire, Calman-Levy, Paris.

Portelli, H, (1980), Le Socialisme Francais Tel Qu'il Est, PUF, Paris.

Queyranne, J-J, (1982), Les régions et la décentralisation culturelle. Les conventions de développment culturel régional, Documentation Francaise, Paris.

Rocard, M. (1966), 'Décoloniser la Province - Documents des Colloques Régionaux de Grenoble, Correspondence Municipale, ADELS No. 91.

Rocard, M. (1981), 'La région, une idée neuve pour la gauche', Pouvoirs, 19.

Rogers, V. (1984), 'Regional Decentralisation in France : The Case of Brittany', Politics 4.1.

Roquette, Y. (1982), 'Dans la Décentralisation Y Aura-t-il Place Pour les Nations?' Après-Demain, 240.

Sadran, P. (1982), 'La Régionalisation Française en Pratique. Esquisse d'un Bilan, 1972-80, in Mény, Y, (ed.), Dix Ans de Régionalisation en Europe, Cujas, Paris.

Shaw, E. (1983), 'French Socialism in the 1980s', in Bel;l, D. and Shaw, E. (eds.) The Left in France, Spokesman, Nottingham

Smith, A.D. (1981), The Ethnic Revival, Cambridge University Press, Cambridge.

Stephens, M. (1976), Linguistic Minorities in Western Europe, Gomer, Llandysul.

Tarrow, S. (1978), 'Regional Policy, Ideology and Peripheral Defense. The Case of Fos-sur-mer', In Tarrow, S., Katzenstein, P.J. and Graziano, L. (eds.) Territorial Politics in Industrial Nations, Praeger, New York.

Thoenig, J-C (1979), 'Local Government Institutions and the Contemporary Evolution of French Society', in Lagroye, J and Wright, V. (eds.) Local Government in Britain and France, Allen and Unwin, London.

Touraine, A., Dubet, F., Hegedus, Z., and Wieviorka, M. (1981), Le Pays Contre l'Etat. Luttes Occitanes, Seuil, Paris.

Vié, J-E, (1982), Du préfet au commissaire de la republique ou le gouvernment de l'imposture', Le Monde, 1 July.

Villeneuve, B. and de Virieu, F-H, (1981), Le Nouveau Pouvoir, J-C Lattes, Paris.

Weber, E, (1977), Peasants into Frenchmen. The Modernisation of Rural France, 1870-1914, Chatto and Windus, London .

Worms, J-P, (1980), 'La Décentralisation: Une Strategie Socialiste de Changement Sociale,' Recherche Sociale, 75.

Wright, V. (1978), The Government and Politics of France, Hutchinson, London.

Wright, V. (1979), in Lagroye, J., and Wright, V. Local Government in Britain and France, Allen and Unwin, London.

Wright, V. and Machin, H. (1975), 'The French Regional Reforms of July 1972: a Case of Disguised Centralisation?', Policy and Politics, III.3.

Index

agriculture 95
Alfonsi, Prosper 100, 102
aménagement du territoire 22, 24, 27, 65, 114, 112
arrondissements 77-9
Association of Mayors 73
autogestion 32, 55-6, 58, 75, 90

Barre, Raymond 121
Basque country 94
Bou, tibert 96
Britanny 38, 40-4
Bucchini, Dominique 101
CELIB 36, 412-2, 59
CERES 115, 131
CFDT 55, 131
CGT 41, 55, 127
Chevènement, Jean-Pierre 131
Chirac, Jacques 29, 78, 100
CODERS 24, 25, 41-2
Commissaire de la République 17, 21, 72, 73, 74, 80, 81, 83, 106, 118, 134
Communist Party (see PCF)
Conseil d'État 113
Corsica 92-108
Crépeau, Michel 61
Cresson, Edith 80
culture 82
cumul des mandats 25, 67, 74, 75, 83, 89, 91, 106, 108, 129, 134

DATAR 26, 37, 112, 116, 128
Debarge, Marcel 74
Debré, Michel 106

Defferre, Gaston, 3, 9, 15, 57, 70, 77, 80, 97, 101, 133
Dotation Globale de Décentralisation 86
Dotation Globale d'Équipement 85
Dotation Globale de Fonctionnement 85

éducation 82
electoral system 76-7
European Community 115, 127
external services 73

Félibrige 45, 46, 50
Ferry, Jules 6
Fifth Republic 19, 23-7
finance 83-7
FLNC 96, 99, 103
Fourth Republic 36

Giscard d'Estaing, Valéry 29, 96, 97, 109, 110, 121
Grenoble 113
Guichard, Olivier 30-1, 71
general competence 79
grands corps 11

housing 80, 81

IGAMES 22

jacobins 5, 8, 16, 79
Joxe, Pierre 133

Lang, Jack 80

Languedoc-Roussillon 119-20 126-7
Leccia, Bastien 98,103
Lille 83
Lorraine 120
Lutte occitane 57-8

Mauroy, Pierre 9, 65
mayor 75, 81, 96
MCA 105
Mendès-France, Pierre 22
Ministry of the Interior 8, 72, 118
Mistral, Fréderic 36
Mitterrand, Francois 3, 10, 15, 33, 89, 102, 104
MRG 61, 100, 101, 102

Napoléon I 16, 93
Napoléon III 17
National Front 105
Nord-Pas de Calais 121, 126
notables 11, 12, 13, 18, 25, 50, 52, 60, 67, 71, 83, 89, 106

officials of local government 87-9

Pantaloni, Ange 101
Paoli, Pascal 93
Paris-Lyon-Marseille law 75, 77-9
Parti Socialiste 13, 51, 56, 62, 63-6, 74, 77, 97, 99, 100, 109, 113, 130
PCF (Communist Party) 101, 104, 111, 127
permis de construire 81
Phlipponneau, Michel 42, 59
Plan, National 67, 90, 109 121-7
Pompidou, Georges 28, 29
prefects 7, 9, 16, 19, 24, 73
proportional representation 105

Questiaux, Nicole 80
Quilliot, Roger 80

Rigout, Marcel 80
Rocca-Serra, Jean-Paul de 100, 102, 106
Rocard, Michel 40, 111, 121
Rouquette, Yves 49
Rousseau, Jean-Jacques 93

SAC 98
saupoudrage 74, 90, 126, 132
Savary, Alain 57, 80, 131
Senate 27, 71
SFIO 55, 56, 59
social services 81
Third Republic 17,18, 19, 20 41, 54
tourism 94
Tresorier-Payeurs Generaux 71
tutelle 7, 8, 71, 72, 73, 74 82, 83, 84, 86

UDF 100
Union de la Gauche 76
UPC 99, 103, 105

Vichy regime 20-1
Vivre Ensemble 30, 71
Volem Viure at Pais 46-8, 49, 127